Energy, Environment, and Sustainability

Series editors

Avinash Kumar Agarwal, Department of Mechanical Engineering, Indian Institute
of Technology Kanpur, Kanpur, Uttar Pradesh, India
Ashok Pandey, Distinguished Scientist, CSIR-Indian Institute of Toxicology
Research, Lucknow, Uttar Pradesh, India

This books series publishes cutting edge monographs and professional books focused on all aspects of energy and environmental sustainability, especially as it relates to energy concerns. The Series is published in partnership with the International Society for Energy, Environment, and Sustainability. The books in these series are editor or authored by top researchers and professional across the globe. The series aims at publishing state-of-the-art research and development in areas including, but not limited to:

- Renewable Energy
- Alternative Fuels
- Engines and Locomotives
- Combustion and Propulsion
- Fossil Fuels
- Carbon Capture
- Control and Automation for Energy
- Environmental Pollution
- Waste Management
- Transportation Sustainability

More information about this series at http://www.springer.com/series/15901

Narasinha Shurpali · Avinash Kumar Agarwal
V. K. Srivastava
Editors

Greenhouse Gas Emissions

Challenges, Technologies and Solutions

Editors
Narasinha Shurpali
Department of Environmental
  and Biological Sciences
University of Eastern Finland
Kuopio, Finland

V. K. Srivastava
Sankalchand Patel University
Visnagar, Gujarat, India

Avinash Kumar Agarwal
Department of Mechanical Engineering
Indian Institute of Technology Kanpur
Kanpur, Uttar Pradesh, India

ISSN 2522-8366   ISSN 2522-8374 (electronic)
Energy, Environment, and Sustainability
ISBN 978-981-13-3271-5   ISBN 978-981-13-3272-2 (eBook)
https://doi.org/10.1007/978-981-13-3272-2

Library of Congress Control Number: 2018961223

© Springer Nature Singapore Pte Ltd. 2019
This work is subject to copyright. All rights are reserved by the Publisher, whether the whole or part
of the material is concerned, specifically the rights of translation, reprinting, reuse of illustrations,
recitation, broadcasting, reproduction on microfilms or in any other physical way, and transmission
or information storage and retrieval, electronic adaptation, computer software, or by similar or dissimilar
methodology now known or hereafter developed.
The use of general descriptive names, registered names, trademarks, service marks, etc. in this
publication does not imply, even in the absence of a specific statement, that such names are exempt from
the relevant protective laws and regulations and therefore free for general use.
The publisher, the authors and the editors are safe to assume that the advice and information in this
book are believed to be true and accurate at the date of publication. Neither the publisher nor the
authors or the editors give a warranty, express or implied, with respect to the material contained herein or
for any errors or omissions that may have been made. The publisher remains neutral with regard to
jurisdictional claims in published maps and institutional affiliations.

This Springer imprint is published by the registered company Springer Nature Singapore Pte Ltd.
The registered company address is: 152 Beach Road, #21-01/04 Gateway East, Singapore 189721,
Singapore

Preface

Energy demand has been rising remarkably due to increasing population and urbanization. Global economy and society are significantly dependent on the energy availability because it touches every facet of human life and its activities. Transportation and power generation are two major examples. Without the transportation by millions of personalized and mass transport vehicles and availability of 24×7 power, human civilization would not have reached contemporary living standards.

The International Society for Energy, Environment and Sustainability (ISEES) was founded at Indian Institute of Technology Kanpur (IIT Kanpur), India, in January 2014 with the aim of spreading knowledge/awareness and catalysing research activities in the fields of energy, environment, sustainability and combustion. The society's goal is to contribute to the development of clean, affordable and secure energy resources and a sustainable environment for the society and to spread knowledge in the above-mentioned areas and create awareness about the environmental challenges, which the world is facing today. The unique way adopted by the society was to break the conventional silos of specializations (engineering, science, environment, agriculture, biotechnology, materials, fuels, etc.) to tackle the problems related to energy, environment and sustainability in a holistic manner. This is quite evident by the participation of experts from all fields to resolve these issues. ISEES is involved in various activities such as conducting workshops, seminars and conferences in the domains of its interest. The society also recognizes the outstanding works done by the young scientists and engineers for their contributions in these fields by conferring them awards under various categories.

The second international conference on "Sustainable Energy and Environmental Challenges" (SEEC-2018) was organized under the auspices of ISEES from 31 December 2017 to 3 January 2018 at J N Tata Auditorium, Indian Institute of Science Bangalore. This conference provided a platform for discussions between eminent scientists and engineers from various countries including India, USA, South Korea, Norway, Finland, Malaysia, Austria, Saudi Arabia and Australia. In this conference, eminent speakers from all over the world presented their views related to different aspects of energy, combustion, emissions and alternative energy resources for sustainable development and a cleaner environment. The conference presented five high-voltage plenary talks from globally renowned experts on topical themes, namely "Is It Really the End of Combustion Engines and Petroleum?" by Prof. Gautam Kalghatgi, Saudi Aramco; "Energy Sustainability in India: Challenges and Opportunities" by Prof. Baldev Raj, NIAS Bangalore; "Methanol Economy: An Option for Sustainable Energy and Environmental Challenges" by Dr. Vijay Kumar Saraswat, Hon. Member (S&T), NITI Aayog, Government of India; "Supercritical Carbon Dioxide Brayton Cycle for Power Generation" by Prof. Pradip Dutta, IISc Bangalore; and "Role of Nuclear Fusion for Environmental Sustainability of Energy in Future" by Prof. J. S. Rao, Altair Engineering.

The conference included 27 technical sessions on topics related to energy and environmental sustainability including 5 plenary talks, 40 keynote talks and 18 invited talks from prominent scientists, in addition to 142 contributed talks, and 74 poster presentations by students and researchers. The technical sessions in the conference included Advances in IC Engines: SI Engines, Solar Energy: Storage, Fundamentals of Combustion, Environmental Protection and Sustainability, Environmental Biotechnology, Coal and Biomass Combustion/Gasification, Air Pollution and Control, Biomass to Fuels/Chemicals: Clean Fuels, Advances in IC Engines: CI Engines, Solar Energy: Performance, Biomass to Fuels/Chemicals: Production, Advances in IC Engines: Fuels, Energy Sustainability, Environmental Biotechnology, Atomization and Sprays, Combustion/Gas Turbines/Fluid Flow/Sprays, Biomass to Fuels/Chemicals, Advances in IC Engines: New Concepts, Energy Sustainability, Waste to Wealth, Conventional and Alternate Fuels, Solar Energy, Wastewater Remediation and Air Pollution. One of the highlights of the conference was the rapid-fire poster sessions in (i) Energy Engineering, (ii) Environment and Sustainability and (iii) Biotechnology, where more than 75 students participated with great enthusiasm and won many prizes in a fiercely competitive environment. More than 200 participants and speakers attended this four-day conference, which also hosted Dr. Vijay Kumar Saraswat, Hon. Member (S&T), NITI Aayog, Government of India, as the chief guest for the book release ceremony, where 16 ISEES books published by Springer under a special dedicated series "Energy, Environment, and Sustainability" were released. This is the first time that such significant and high-quality outcome has been achieved by any society in India. The conference concluded with a panel discussion on "Challenges, Opportunities & Directions for Future Transportation Systems", where the panellists were Prof. Gautam Kalghatgi, Saudi Aramco; Dr. Ravi Prashanth, Caterpillar Inc.; Dr. Shankar Venugopal, Mahindra and Mahindra; Dr. Bharat Bhargava, DG, ONGC Energy Centre; and Dr. Umamaheshwar, GE Transportation, Bangalore. The panel discussion was moderated by Prof. Ashok Pandey, Chairman, ISEES. This conference laid out the road map for technology development, opportunities and challenges in energy, environment and sustainability domains. All these topics are very relevant for the country and the world in the present context. We acknowledge the support received from various funding agencies and organizations for the successful conduct of the second ISEES conference SEEC-2018, where these books germinated. We would therefore like to acknowledge SERB, Government of India (special thanks to Dr. Rajeev Sharma, Secretary); ONGC Energy Centre (special thanks to Dr. Bharat Bhargava); TAFE (special thanks to Sh. Anadrao Patil); Caterpillar (special thanks to Dr Ravi Prashanth); Progress Rail, TSI, India (special thanks to Dr. Deepak Sharma); Tesscorn, India (special thanks to Sh. Satyanarayana); GAIL, Volvo; and our publishing partner Springer (special thanks to Swati Meherishi).

The editors would like to express their sincere gratitude to a large number of authors from all over the world for submitting their high-quality work in a timely manner and revising it appropriately at short notice. We would like to express our special thanks to Drs. Matthew Bell, Mokhele Moelisti, Eleanora Nistor, Kofi Boateng, Beibei Yan and Rafael Eufrasio, who reviewed various chapters of this book and provided very valuable suggestions to the authors to improve their manuscript.

Climate change is a global threat. The impacts of a changing climate are evident in the form of extreme climate, increased droughts and floods, sea level rise and permafrost thawing in the Arctic with an overall impact on the global radiation balance and surface temperature. Our dependence on fossil fuels, deforestation and land use are the major anthropogenic causes that have lead to the changes in our atmospheric composition. Greenhouse gases (GHGs), essential for maintaining an average global temperature, have seen a rise in their concentrations in the atmosphere, owing to the human perturbation of the climate system. To curb the unprecedented rise in atmospheric GHGs, the global community is striving to account for the various sources and sinks of GHGs, so that high emitters can be identified, opportunities to mitigate climate change can be formulated, and adaptation majors can be sorted out. In this context, this book serves to present case studies on GHG emission scenarios from different parts of the world.

Kuopio, Finland
Kanpur, India
Visnagar, India

Narasinha Shurpali
Avinash Kumar Agarwal
V. K. Srivastava

Contents

1 Introduction to Greenhouse Gas Emissions 1
Narasinha Shurpali, A. K. Agarwal and V. K. Srivastava

2 Greenhouse Gas Fluxes of Agricultural Soils in Finland 7
Kristiina Regina, Jaakko Heikkinen and Marja Maljanen

3 Greenhouse Gas Exchange from Agriculture in Italy 23
Anna Dalla Marta and Leonardo Verdi

4 GHG Emissions and Mitigation in Romanian Vineyards 33
Eleonora Nistor, Alina Georgeta Dobrei, Alin Dobrei
and Narasinha Shurpali

**5 Agricultural Cropping Systems in South Africa
and Their Greenhouse Gas Emissions: A Review** 57
Mphethe Tongwane, Sewela Malaka and Mokhele Moeletsi

6 Agricultural Greenhouse Gases from Sub-Saharan Africa 73
Kofi K. Boateng, George Y. Obeng and Ebenezer Mensah

**7 The UK Path and the Role of NETs to Achieve
Decarbonisation** 87
Rafael M. Eufrasio-Espinosa and S. C. Lenny Koh

**8 Measuring Enteric Methane Emissions from Individual
Ruminant Animals in Their Natural Environment** 111
Matt J. Bell

9 Crop Residue Burning: Effects on Environment 127
Ritu Mathur and V. K. Srivastava

10 Rooftop Solar Power Generation: An Opportunity to Reduce Greenhouse Gas Emissions 141
Neeru Bansal, V. K. Srivastava and Juzer Kheraluwala

11 Renewable Energy in India: Policies to Reduce Greenhouse Gas Emissions 161
Neeru Bansal, V. K. Srivastava and Juzer Kheraluwala

Editors and Contributors

About the Editors

Narasinha Shurpali is a senior researcher in the Department of Environmental and Biological Sciences, University of Eastern Finland (UEF), Kuopio Campus. He received his M.Sc. and Ph.D. from the College of Agriculture, Mahatma Phule Agricultural University (India) and University of Nebraska (USA), respectively, and subsequently worked at various international institutes including the University of Antwerp (Belgium), Indian Council of Agricultural Research (India), University of Indiana (USA) and the Finnish Forest Research Institute (Finland). His primary research interest is in using measurement techniques to understand the carbon and nitrogen biogeochemical cycles in agriculture and other complex ecosystems such as natural and managed peatlands, forests on organic and mineral soils, arctic ecosystems, and simulation modelling of biogeochemical cycles of C and N using field data to validate the models, which represent an important tool in enhancing our understanding of the atmosphere–biosphere exchange.

Avinash Kumar Agarwal is a professor in the Department of Mechanical Engineering at Indian Institute of Technology Kanpur. His areas of interest are IC engines, combustion, alternative fuels, conventional fuels, optical diagnostics, laser ignition, HCCI, emission and particulate control, and large bore engines. He has published 24 books and more than 230 international journal and conference papers. He is a fellow of SAE (2012), ASME (2013), ISEES (2015) and INAE (2015). He has received several awards such as the prestigious Shanti Swarup Bhatnagar Award in engineering sciences (2016), Rajib Goyal Prize (2015) and NASI-Reliance Industries Platinum Jubilee Award (2012).

V. K. Srivastava is Provost (Vice Chancellor) at Sankalchand Patel University, Visnagar, Gujarat (India). His area of research interest includes waste water treatment, solid waste management, Environmental issues of industries and energy recovery by using plasma technology. He is peer reviewer of large number of research papers being published in reputed high indexed National and International Journals in above mentioned research areas. He has published and presented more than 50 research papers in National and International Journals and Conferences and also number of books and chapters with National and International publishers. He has worked on many government- and industry-sponsored projects and is a member of many technical and professional societies including the European Geosciences Union, Society of Environment Toxicology and Chemistry, American Chemical Society and Royal Society of Chemistry.

Contributors

A. K. Agarwal Department of Mechanical Engineering, Indian Institute of Technology Kanpur, Kanpur, Uttar Pradesh, India

Neeru Bansal CEPT University, Ahmedabad, Gujarat, India

Matt J. Bell University of Nottingham, Leicestershire, UK

Editors and Contributors

Kofi K. Boateng Department of Agricultural and Biosystems Engineering, Kwame Nkrumah University of Science and Technology, UPO, KNUST, Kumasi, Ghana

Alin Dobrei Banat University of Agricultural Sciences and Veterinary Medicine "King Michael I of Romania", Timişoara, Romania

Alina Georgeta Dobrei Banat University of Agricultural Sciences and Veterinary Medicine "King Michael I of Romania", Timişoara, Romania

Rafael M. Eufrasio-Espinosa Advanced Resource Efficiency Centre (AREC), The University of Sheffield, Management School, Sheffield, UK

Jaakko Heikkinen Natural Resources Institute Finland (Luke), Jokioinen, Finland

Juzer Kheraluwala Ernst & Young LLP, New Delhi, India

S. C. Lenny Koh Advanced Resource Efficiency Centre (AREC), The University of Sheffield, Management School, Sheffield, UK

Sewela Malaka Agricultural Research Council – Institute for Soil, Climate and Water, Pretoria, South Africa

Marja Maljanen Department of Environmental and Biological Sciences, University of Eastern Finland, Kuopio, Finland

Anna Dalla Marta Department of Agrifood Productions and Environmental Sciences, University of Florence, Florence, Italy

Ritu Mathur Government R.R. Autonomous College, Alwar, Rajasthan, India

Ebenezer Mensah Department of Agricultural and Biosystems Engineering, Kwame Nkrumah University of Science and Technology, UPO, KNUST, Kumasi, Ghana

Mokhele Moeletsi Agricultural Research Council – Institute for Soil, Climate and Water, Pretoria, South Africa; Risk and Vulnerability Assessment Centre, University of Limpopo, Sovenga, South Africa

Eleonora Nistor Banat University of Agricultural Sciences and Veterinary Medicine "King Michael I of Romania", Timişoara, Romania

George Y. Obeng Technology Consultancy Center and Mechanical Engineering Department, College of Engineering, Kwame Nkrumah University of Science and Technology, UPO, KNUST, Kumasi, Ghana

Kristiina Regina Natural Resources Institute Finland (Luke), Jokioinen, Finland

Narasinha Shurpali Department of Environmental and Biological Sciences, Kuopio, Finland

V. K. Srivastava Sankalchand Patel University, Visnagar, Gujarat, India

Mphethe Tongwane Agricultural Research Council – Institute for Soil, Climate and Water, Pretoria, South Africa

Leonardo Verdi Department of Agrifood Productions and Environmental Sciences, University of Florence, Florence, Italy

Chapter 1
Introduction to Greenhouse Gas Emissions

Narasinha Shurpali, A. K. Agarwal and V. K. Srivastava

Abstract Climate change is a global threat. The increasing concentrations of greenhouse gases (GHGs—carbon dioxide, methane, and nitrous oxide) in the atmosphere are blamed for the changing global climate. The greenhouse gases play a key role in causing climate change, and the biosphere can contribute positively as well as negatively to the atmospheric GHG concentrations through feedback processes. Knowledge of the feedback processes and their interactions with climate change and human activities is necessary if we are to understand to what impact feedback processes will have on climate change and to what extent manipulation of the biosphere will actually have the desired beneficial effects. In this context, a significant amount of scientific research work has been done and is continuing across different parts of the world to characterize the sources and sinks of GHGs. Thus, the book entitled, 'Greenhouse Gas Emissions,' is a compilation of select case studies on the topic authored by international scientists from different parts of the world.

Keywords Climate change · Agriculture · Decarbonisation · Ruminants
Crop residue burning · Solar power · Mitigation

N. Shurpali (✉)
Department of Environmental and Biological Sciences, Yliopistoranta 1 DE,
PO Box 1627, 70211 Kuopio, Finland
e-mail: narasinha.shurpali@uef.fi

A. K. Agarwal
Department of Mechanical Engineering, Indian Institute of Technology Kanpur,
Kanpur 208016, Uttar Pradesh, India
e-mail: akag@iitk.ac.in

V. K. Srivastava
Sankalchand Patel University, Visnagar, Gujarat, India
e-mail: drvks9@gmail.com

© Springer Nature Singapore Pte Ltd. 2019
N. Shurpali et al. (eds.), *Greenhouse Gas Emissions*, Energy, Environment,
and Sustainability, https://doi.org/10.1007/978-981-13-3272-2_1

The atmosphere forms a major part of our environment. The life on Earth dynamically responds to this environment. The atmosphere interacts with the biosphere, hydrosphere, cryosphere, and lithosphere on timescales from seconds to millennia (Jerez et al. 2018) and on spatial scales from molecules to the global level. Changes in one component are directly or indirectly communicated to the other components through complex processes and feedbacks. Human and societal actions, such as energy and land use and various natural feedback mechanisms involving the biosphere and atmosphere, have major impacts on the complex interplay between radiatively important trace gases in the atmosphere and climate (Dai 2016). The carbon dioxide content of the atmosphere has increased by 43% since 1750 (Ciais et al. 2013). The atmospheric CO_2 concentration as measured at the Mauna Loa laboratory during June 2018 was 411 ppm (as opposed to about 280 ppm during the preindustrial times). The growth rate of atmospheric CO_2 during the 1960–69 decade was 0.85 ppm $year^{-1}$, while it climbed to 2.28 ppm $year^{-1}$ during the recent 10-year (2008–17) period. The reason for such a drastic change in the atmospheric composition is attributed to our overdependence on fossil fuels for energy and deforestation. Corresponding to this rise in CO_2 content, the mean global surface temperature has already increased by 0.85 °C, compared to the preindustrial era (Hansen et al. 2010). Additionally, global methane concentrations have increased from 722 parts per billion (ppb) in preindustrial times to 1834 ppb by 2013, an increase by a factor of 2.5 and the highest value in at least 800,000 years. Nitrous oxide (N_2O) is among the most important greenhouse gases (GHGs) as its one molecule has about 300 times greater warming potential than that of CO_2 over a 100-year time horizon (Myhre et al. 2013). It is produced both in natural and managed soils, agricultural soils being the largest anthropogenic source. Changes in the atmospheric composition of these GHGs are causes for a changing climate across the globe. The impacts of climate change are becoming evident across all continents in the warmer oceans, reduced snow and ice cover and rising sea levels. With this in view, we have made an attempt in this book to gather information on GHG dynamics in different parts of the world and present a few case studies on the possibilities for mitigation of climate change.

Agriculture in northern European regions, such as Finland, is limited by the short growing seasons and low cumulative degree days during the growing period. Climate change is projected to lengthen the growing seasons and increase the growing degree days. Crop yields are projected to increase in Northern Europe, although the projections allow for both positive and negative impact on crop yields. Finland is a northern country with cool and temperate climate. This has implications for the greenhouse gas balance of cultivated soils. Utilizing organic soils for food production is unavoidable in Finland owing to its high coverage of peat soils. The greenhouse gas emissions per hectare are several folds on organic soils than on mineral soils. Thus, despite their proportion being only ten percent of the total cultivated area in Finland, the organic soils are a dominant source of agricultural carbon dioxide and nitrous oxide emissions at the national level. Chapter 2 provides an account of the emissions from organic agricultural soils in Finland as relevant to the land use and climate change policies in Finland.

In Italy, in addition to industry, transport, and energy sectors, agriculture is one of the main sources of anthropogenic greenhouse gases (GHGs) emissions. Intensive and extensive cultivation practices such as fertilization and fuel consumption for tractors are the more impactful factors in terms of global GHGs. Italy does not represent an exception to the rule, and owing to its high variability in its environmental and geomorphological conditions, a wide range of agricultural systems are adopted in the country with varying GHGs emission potentials. CO_2 emissions are primarily produced from mechanized agriculture in the country. Northern Italy is known for its paddy cultivation, which is the main source of CH_4 emissions that are produced in flooded fields from anaerobic microorganisms. Crop and animal husbandry are practiced in tandem, and thus, methane emissions from farm animals are also an important GHG source in the country. In addition, the intensive use of fertilizers contributes to N-based emissions following nitrification/denitrification and N-volatilization processes into the soil. Chapter 3 gives an account of how the Italian agriculture plays a key role in GHGs emissions.

Romania is a major European wine country with rich historic and cultural traditions, many of them directly related to wine. The national policies are geared towards making this country a producer of high-quality wine and thus a valued member of the world wine community. The total area under viticulture in Romania represents about 2% of the total arable land area. Viticulture accounts for about 7% of the total agricultural production and wine ranks third among the exported agri-food products. While there are not many studies reporting GHG emissions from grape cultivation in Romania, Chap. 4 provides the perspective on Romanian viticulture.

Agriculture in the sub-Saharan African sub-continent, although still rudimentary in terms of management practices and production efficiency, provides the mainstay for majority of its people. Chapter 5 takes a look at the sub-Saharan African agriculture, its contribution to the emission of Greenhouse gases, and their pathways. It aims to address the effects of a changing climate on SSA agricultural productivity. The contribution of SSA agriculture to the socioeconomic well-being of its people is also discussed. Adaptation and resilience building among the dominating smallholder farmers in the region are captured as well as the factors that hinder the effective scaling up of strategies aimed at ameliorating the effects of climate variability on local agriculture. Finally, policy interventions geared toward the significant reduction of climate change effects on SSA are discussed.

South Africa is a major emitter of greenhouse gases (GHG) and accounts for 65% and 7% of Africa's total emissions and agricultural emissions, respectively. South Africa has a dual agricultural economy, comprising a well-developed commercial sector and subsistence-oriented farming in the rural areas. The country has an intensive management system of agricultural lands. Agriculture, Forestry and Land Use are the second largest emitter in the country. Chapter 6 presents characteristics of GHG emissions from crop management in South Africa with a national perspective sustainable mitigation options.

In 2017, the UK powered itself for a full day without coal for the first time since the Industrial Revolution. In addition, in the beginning of this year, it laid out a strategy to phase out all coal-fired power plants by 2025. This has been made possible by an upsurge in the use of renewable energy in the country. Such efforts to combat climate change show a significant decrease in CO_2 emissions during the last 5 years. Adoption of positive national climate change strategies has lead UK on a steady transition toward a low-carbon economy. Chapter 7 reviews the current state of the greenhouse gas emissions in the UK and describes what measures UK has adopted to take the nation on the path of a low-carbon economy.

While the above chapters provide country-specific GHG emission scenarios, this book also provides insights into other issues that are relevant to national GHG accounting. Some such case studies include methane emissions from cattle and emissions from crop residue burning. Humans depend on livestock as they are an important source of meat, milk, fiber, and labor. Energy is lost in the form of methane gas when the ruminants digest plant material through rumen fermentation. Ruminant livestock is a significant source of atmospheric methane, with an estimated 17% of global enteric methane emissions from livestock. Methane is a potent GHG with about 25 times higher warming potential than CO_2. The chapter on measuring methane emissions from ruminants (Chap. 8) provides a review on the measurement techniques and discusses their advantages and limitations with a perspective on accurate accounting of these emissions from this important source.

India is one of the key global producers of food grain, oilseed, sugarcane, and other agricultural products. Agriculture generates huge amounts of crop residues. With an expected increase in food production in the future, crop residue generation will also increase. These leftover residues exhibit not only resource loss but also a missed opportunity to improve a farmer's income. Currently, the farmers in India resort to residue burning, a practice that is perceived to enhance soil carbon sequestration. While such a practice is being followed since a long time, its impact on the environment is not well understood (Chap. 9). There is a need for extensive research with large-scale GHG measurements from crop residue burning in India.

At the Paris COP21 climate summit held at the end of 2015, a Breakthrough Energy Coalition and Mission Innovation plans were formulated by the participating countries. These are strategies aimed at reducing global GHGs, the use of clean energy and limiting the global surface temperature increase to 2 °C or less by 2050. With abundant solar energy available in India, attempts are in full swing to harness this renewable source of energy in the country. The chapter on rooftop solar power generation (Chap. 10) exemplifies these attempts with a case study from a metropolitan city in western India, while the chapter on renewable energy sources in India (Chap. 11) focusses on policies of the central and state governments in India to promote renewable energy, especially solar energy, to reduce national GHG emissions.

References

Ciais P et al (2013) Climate change 2013: the physical science basis. Contribution of working group I to the fifth assessment report of the intergovernmental panel on climate change. Cambridge University Press

Dai A (2016) Future warming patterns linked to today's climate variability. Sci Rep 6, Article number: 19110

Hansen J, Ruedy R, Sato M, Lo K (2010) Global surface temperature change. Rev Geophys 48, Article number: RG4004

Jerez S, López-Romero JM, Turco M, Jiménez-Guerrero P, Vautard R, Montávez JP (2018) Impact of evolving greenhouse gas forcing on the warming signal in regional climate model experiments. Nat Commun 9, Article number: 1304

Myhre G, Shindell D, Bréon F-M, Collins W, Fuglestvedt J, Huang J, Koch D, Lamarque J-F, Lee D, Mendoza B, Nakajima T, Robock A, Stephens G, Takemura T, Zhang H (2013) Anthropogenic and natural radiative forcing. In: Stocker TF, Qin D, Plattner G-K, Tignor M, Allen SK, Boschung J, Nauels A, Xia Y, Bex V, Midgley PM (eds) Climate change 2013: the physical science basis. Contribution of working group I to the fifth assessment report of the intergovernmental panel on climate change. Cambridge University Press, Cambridge, United Kingdom and New York, NY, USA

Chapter 2
Greenhouse Gas Fluxes of Agricultural Soils in Finland

Kristiina Regina, Jaakko Heikkinen and Marja Maljanen

Abstract Finland is a northern country with cool and humid climate. This has implications for the greenhouse gas balance of cultivated soils. Utilizing organic soils for food production is unavoidable in a country with high coverage of peat soils. As the greenhouse gas emissions per hectare are several folds on organic soils compared to mineral soils, organic soils are a dominant source of agriculture-related carbon dioxide and nitrous oxide emissions in country scale although their proportion is only 10% of the field area. Another factor that exposes fields to high losses of nutrients and organic matter is the short growing season and the resulting long non-vegetated period. The review of existing data shows that emissions of carbon dioxide and nitrous oxide are the most important components of the total greenhouse gas balance, whereas fluxes of methane are negligible in drained cultivated soils. Generally, the total emissions are higher from annual than perennial cropping. Climate and agricultural policies have tightening requirements for all economic sectors, and this imposes new challenges to agricultural management. As soils are a major source of greenhouse gas emissions in agriculture, special attention should be paid on developing mitigation measures and practices that reduce the climatic impact of cultivated soils.

K. Regina (✉) · J. Heikkinen
Natural Resources Institute Finland (Luke), Tietotie 4, 31600 Jokioinen, Finland
e-mail: kristiina.regina@luke.fi

J. Heikkinen
e-mail: jaakko.heikkinen@luke.fi

M. Maljanen
Department of Environmental and Biological Sciences, University of Eastern Finland,
Yliopistonranta 1, 70210 Kuopio, Finland
e-mail: marja.maljanen@uef.fi

© Springer Nature Singapore Pte Ltd. 2019
N. Shurpali et al. (eds.), *Greenhouse Gas Emissions*, Energy, Environment,
and Sustainability, https://doi.org/10.1007/978-981-13-3272-2_2

2.1 Agriculture, Climate and Soil Types in Finland

Finland is located in northern Europe between the northern latitudes 60°–70° and eastern longitudes 20°–30°. It has the northernmost agricultural regions of the European Union. Forest is the dominant land use, and 8% of the land area of the country is used as croplands. Agriculture is concentrated in the southern and western regions of the country (Fig. 2.1). Annual cropping is restricted to the southern parts of the country, whereas the northern regions with short growing season suit best for grasslands and milk production.

Half of the agricultural area is used for cereal cropping and about 30% for grass production (Luke 2016). Grass is mainly cultivated in crop rotations, and permanent grasslands are not common; most grasslands are renewed every 3–4 years due to damages in the sward during the winter.

The conventional and most common management of cropland includes autumn tillage, crop residue retention and the use of mineral fertilizer and pesticides. Manure is typically used as fertilizer on the same farm where it is produced. Liming is needed under Finnish conditions to render the naturally acid soils to suitable for agricultural production. Drainage is essential for maintaining proper cultivation conditions; 60–70% of the croplands currently are equipped with a subsurface drainage system, and the rest has open ditches.

The climate is humid boreal with mean temperature ranging from -2 to 5 °C and annual precipitation from 450 to 750 mm depending on the region. Length of the growing season varies from 105 to 185 days in the different regions of the country, and the effective temperature sums from 600 to 1400 °C.

Due to the northern and humid conditions of Finland, peatlands are characteristic for the landscape. Originally half of the area of Finland has been peat soil. Half of that is not in a pristine state but has been drained for different uses, mainly forestry. A smaller fraction of the peat soils is utilized as croplands, and currently about 10% of the cultivated area is under peat or other organic soils (Table 2.1). The rest is different mineral soils with clay soils dominating in the south and coarse soils in the north.

2.2 Agriculture as a Source of Greenhouse Gas Emissions

Greenhouse gases relevant for agriculture are methane (CH_4), carbon dioxide (CO_2) and nitrous oxide (N_2O) with the two latter constituting most of the emissions from soils. In agriculture, these gases are mainly released from processes where organic matter decomposes or mineral inputs are used as substrates in microbial processes as part of nutrient cycles. Sources of greenhouse gas emissions in agriculture are soils, enteric fermentation of production animals and manure storages. Nutrient cycles are faster in agricultural soils compared to native soils, and this imposes

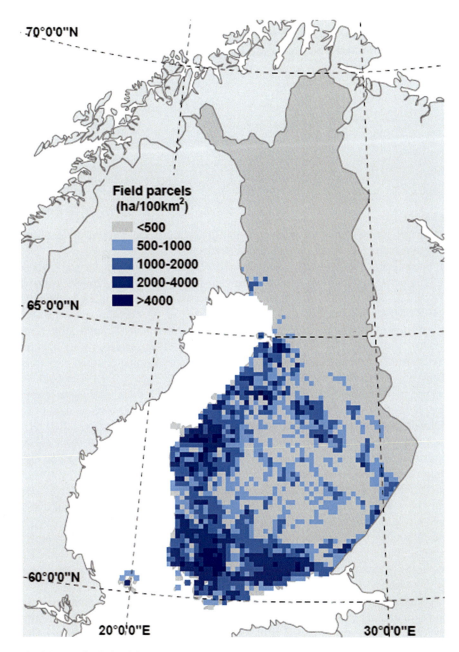

Fig. 2.1 Croplands in Finland

Table 2.1 Soil and cultivation types of cropland in Finland

	% of cropland area
Soil type	
Clay	40.8
Fine	20.6
Coarse	27.6
Organic	11.0
Cropping sequences	
Annual cropland	38.5
Perennial cropland	12.5
Crop rotation	49.0

Reference: Heikkinen et al. (2013)

challenges for managing the production systems in a way that minimizes losses and the environmental impact of agricultural production.

Greenhouse gas emissions are reported annually under the United Nations Framework Convention on Climate Change, and the reports are publicly available (UNFCCC 2018). Total greenhouse gas emissions of Finland varied between 55 and 81 Mt CO_2 eq. in 2006–2016, but only part of agricultural soil emissions is included in this figure, namely nitrous oxide emissions from fertilization, crop residues and decomposition of peat in organic soils. These emissions are reported under sector "Agriculture". Carbon stock changes from mineral soils, and emissions of CO_2 from organic soils are reported in the sector "land use, land-use change and forestry" (LULUCF) which is not part of the share referred to as total emissions. Methane fluxes from soils are not reported as they are of minor significance and not a mandatory category. All agricultural emissions (with LULUCF included) have been 20–25% of the total annual emissions of Finland. Soils are the highest single source of emissions in agriculture, which is mainly related to the high proportion of cultivated organic soils in Finland.

The total area of croplands has been extremely stable for the latest decades. However, due to the development towards larger farm size, there has been reallocation of production from mineral to organic soils as the farms quitting production are located in the southern regions, and the enlarging farms are in the peat-rich regions of western and northern Finland. The proportion of organic soils has increased from 8 to 11% in 1990–2016. This is the main reason for clear increase in greenhouse gas emissions from croplands during the same period.

Climate policies have increasing requirements for different economic sectors, and agricultural policies will need to reflect on those better than before. Agricultural emission sources are small and scattered, and the emissions typically feature high uncertainties due to the biological nature of the processes. A large database of emission measurements is needed to enable the design of effective mitigation measures and verification of mitigation effects. This chapter reviews the available data and underlines the development needs.

2.3 Greenhouse Gas Fluxes from Agricultural Soils

2.3.1 Mineral Soils

2.3.1.1 Carbon Dioxide

The balance between carbon input and decomposition of organic matter determines if a field is a source or a sink of carbon. There is only one study that reports net ecosystem exchange of a field on mineral soil (Lind et al. 2016) and that represents cultivation of reed canary grass which is not an especially common crop. Thus, the available data does not allow for estimation of a full carbon balance of a typical field based on measurements (Table 2.2). Evidence from a 35-year field monitoring points to the direction that cultivated mineral soils on the average lose carbon at an annual rate about 200 kg/ha (Heikkinen et al. 2013). The estimate was based on monitoring of about 500 fields and soil sampling only to 15 cm; thus, it represents changes in the topsoil only. The authors deduced that the declining trend is related to warming climate, changes in cropping (less annual crops and varieties with less crop residues) and the young age of fields that may be still losing carbon from the phase of the preceding land use (forest).

The amount of carbon input depends on choices made in cultivation practices. Decomposition is mainly driven by the climatic conditions although, e.g. tillage practices have a role in that as well. Despite the cool climate restricting decomposition of organic matter, it is likely that conventional agricultural practices result in loss of carbon from soil in Finnish conditions. A long-term field experiment in the neighbouring country, Sweden, showed that returning only crop residues with

Table 2.2 Annual greenhouse gas fluxes of cultivated mineral soils

	Mean (g m^2 year^{-1})	Min	Max	n	Refs.
Annual crop					
Net CO_2 exchange	–	–	–		
C loss as yield (CO_2)	–	–	–		
CH_4 flux	-0.04 ± 0.07	-0.12	0.06	7	1; 2
N_2O flux	0.57 ± 0.23	0.20	1.02	31	1; 3; 4; 5
Perennial crop					
Net CO_2 exchange[a]	-950	-961	-939	2	6
C loss as yield (CO_2)	1183	1228	1137	2	6
CH_4 flux	-0.05 ± 0.03	-0.09	0.03	14	1; 2; 10
N_2O flux	0.43 ± 0.31	0.06	1.13	20	1; 3; 4; 7; 8; 9

n = number of annual flux estimates
[a]Negative value = carbon sequestration, positive value = carbon loss
References: 1 (Syvasalo et al. 2006); 2 (Regina et al. 2007); 3 (Syvasalo et al. 2004); 4 (Petersen et al. 2006); 5 (Sheehy et al. 2013); 6 (Lind et al. 2016); 7 (Regina et al. 2006); 8 (Maljanen et al. 2009); 9 (Virkajarvi et al. 2010); 10 (Maljanen et al. 2012b)

no other amendments usually results in the decline of the carbon stock (Katterer et al. 2011). Plant breeding tends to develop varieties with less and less crop residues, which may also complicate maintaining the carbon content of cultivated soils. However, the amount of above-ground plant litter does not seem to be crucial for maintaining the carbon stocks of cultivated soils as the removal or burning of straw did not have an effect in a 30-year field experiment in Southern Finland (Singh et al. 2015).

Converting native ecosystems to agricultural use typically reduces the carbon stock by 20–40%, and the loss of carbon from the soil profile is fastest during the first decades (Guo and Gifford 2002; Karhu et al. 2011). Reaching a new steady state where the carbon input and its loss are in balance may take several decades, and thus it is impossible to say how much of the observed carbon loss is due to the land use change and how much is caused by agricultural management.

2.3.1.2 Methane

Methane is produced microbially in anaerobic conditions and consumed in aerobic conditions. In soils, the conditions can vary from anaerobic to aerobic in time or space. The sites of CH_4 production are the lower soil layers with low oxygen content or soil aggregates favouring anaerobic bacteria. Sites of CH_4 consumption are the topsoil or macropores of the soil. Soil micro- or macroporosity was found to affect the observed rates of CH_4 flux in Finnish clay and sandy soils (Regina et al. 2007). In cultivated soils, the annual balance of CH_4 is usually close to zero most often resulting in more CH_4 being consumed than produced. Emissions of CH_4 have been reported to occur occasionally in wet conditions (Regina et al. 2007), but even then, the annual balance typically indicates net consumption of CH_4. Compared to CO_2, the carbon flows related to CH_4 are minor (Table 2.2). The annual fluxes have ranged from -0.12 to 0.06 g m^{-2} with no clear differences between annual and perennial cropping can be seen. Grazing has been found to change pastures from sink to source of CH_4 emissions due to CH_4 released from the deposited dung (Maljanen et al. 2012b).

2.3.1.3 Nitrous Oxide

In cultivated mineral soils, the most significant gas in the total greenhouse gas budget is N_2O. Average annual emissions of N_2O have been 0.6 g m^{-2} for annual crops and 0.4 g m^{-2} for perennial crops including mostly grass leys (Table 2.2). Annual emissions of N_2O are typically slightly higher from annual cultivation compared to perennial despite the higher fertilization rates on perennial ley production (Regina et al. 2013).

Perennial crops take up nutrients clearly for a longer period annually compared to annual crops and that reduces the amount of nitrogen available for the microbes during the non-vegetated period. The emissions during the period between harvest and sowing represent about 40% of the annual budget of N_2O (Regina et al. 2013) which highlights the importance of the residual nitrogen after harvest in the absence of nutrient uptake of plants. The difference between perennial and annual crops may thus be emphasized in the northern conditions with short growing season of cash crops.

It has been found in many studies that emissions of N_2O are not ceased in the winter time even when the soil is frozen. Availability of nitrate is always good in cultivated soils, and the low oxygen content favours denitrifying bacteria that can be active in microsites with unfrozen water of frozen soil (Teepe et al. 2004). One reason for the high N_2O production at low temperatures can be that N_2O reductase enzymatic activity is inhibited (Muller et al. 2003), and therefore the end product of denitrification is N_2O instead of N_2.

The timing of freezing and soil water content have important effects on the emission of N_2O (Maljanen et al. 2009; Teepe et al. 2004). Also the depth and timing of snow cover can affect N_2O emissions. Snow manipulation experiments have shown that thinner snow cover can lower soil temperatures and increase the extent and duration of soil frost (Maljanen et al. 2009). In frozen soil, N_2O is still produced and accumulated in soil, and it is then rapidly released during thawing (Koponen and Martikainen 2004; Maljanen et al. 2007a, 2009). The N_2O production in frozen soil does not correlate well with the N_2O emitted from soil as a result of the low gas diffusion rate. Therefore, the release of N_2O during winter does not give the correct estimate of N_2O production activity during the winter.

Fertilization rate, especially the amount of mineral nitrogen, has been found to affect the annual emissions when studied in subsets of annual and perennial cropping (Regina et al. 2013). The available data does not allow reliable estimates of the effects of fertilizer type (mineral/organic) on N_2O emissions. Recent evidence shows that external nitrogen inputs induce also emissions of nitric oxide (NO) and gaseous nitrous acid (HONO) that are not greenhouse gases but reactive in the atmosphere (Bhattarai et al. 2018; Maljanen et al. 2007b).

There is some evidence that no-till management increases N_2O emissions from cultivated soils (Sheehy et al. 2013). The increase is related to the more dense structure of the soil and thus higher soil moisture favouring denitrification.

A probable but poorly known hotspot of N_2O emissions are fields on acid sulphate soils. They are located on the former sea bottom of the coastal regions and have large amounts of organic matter in the subsoil due to sedimented materials. The field area on acid sulphate soils is in the range of 43 000–130 000 ha (Yli-Halla et al. 1999). They are characterized by a large stock of nitrogen and high microbial activity that may induce extremely high emissions of N_2O, even of the magnitude of tens of kilograms per hectare when drained for agriculture (Simek et al. 2011, 2014; Petersen et al. 2012; Denmead et al. 2010).

2.3.2 Organic Soils

2.3.2.1 Carbon Dioxide

Losses of carbon from organic soils are typically several folds compared to carbon stock changes in mineral soils. Typically 0.5–2 cm of peat is lost from the topsoil of cultivated soils due to peat decomposition annually (Gronlund et al. 2008) and that represents carbon loss of several tonnes per hectare. Although carbon exchange between the soil and atmosphere forms the majority of the climatic impact of cultivated organic soils, full carbon balance estimates are still rare. The annual net ecosystem exchange has varied between -800 and 3000 g m^{-2} in Finnish studies (Table 2.3). As the reported values show, even in organic soils photosynthesis may sometimes exceed carbon loss from the soil, at least in the case of crops with large biomass. The consideration of the climatic impact must, however, include the biomass transported from the field, and in most cases, this turns the field to net source of carbon even if photosynthesis was able to counteract peat decomposition.

Table 2.3 Annual greenhouse gas fluxes of cultivated organic soils

	Mean (g m^2 year^{-1})	Min	Max	n	GWP (t CO$_2$ eq. ha^{-1} year^{-1})	Refs.
Annual crop						
Net CO$_2$ exchange*	2080 ± 1150	770	3040	4	20.8	1, 2, 3
C loss as yield (CO$_2$)	600 ± 180	460	855	4	6.0	1, 2, 3
CH$_4$ flux	−0.06 ± 0.24	−0.49	0.51	10	−0.02	3, 4, 5
N$_2$O flux	1.74 ± 0.92	0.84	3.79	11	5.2	3, 6, 7
Total					32.0	
Perennial crop						
Net CO$_2$ exchange*	560 ± 1210	−780	2750	8	5.6	1, 2, 3
C loss as yield (CO$_2$)	920 ± 400	280	1570	8	9.2	1, 2, 3
CH$_4$ flux	0.15 ± 0.34	−0.25	0.91	14	0.05	3, 4, 5, 8
N$_2$O flux	1.14 ± 1.47	0.04	5.47	19	3.4	3, 6, 7, 8, 9, 10
Total					18.3	

n = number of annual flux estimates
GWP = global warming potential
*Negative value = carbon sequestration, positive value = carbon loss
References: 1 (Maljanen et al. 2001); 2 (Lohila et al. 2004); 3 (Maljanen et al. 2004); 4 (Maljanen et al. 2003a); 5 (Regina et al. 2007); 6 (Maljanen et al. 2003b); 7 (Regina et al. 2004); 8 (Maljanen et al. 2009); 9 (Maljanen et al. 2010b); 10 (Shurpali et al. 2009)

The existing results suggest lower carbon losses from soils under a perennial than annual crop. This is related to the less frequent disturbance of the soil as well as higher carbon input to the soil, especially from high-yielding grass crops.

2.3.2.2 Methane

Similar to mineral soils, fluxes of CH_4 are close to zero also in drained organic soils (Table 2.3). The mean value for annual crops is slightly negative indicating oxidation of CH_4 in the topsoil, whereas it seems that net production of CH_4 is more common in the denser and moister soil under perennial crops. Periods of wet soil conditions have been found to increase the net flux occasionally as reported, e.g. by Maljanen et al. (2013).

Drainage level and functioning of the drains affect soil moisture status and thus largely determine the flux rates of CH_4. In two fields with similar cultivation practices, the field with poorly functioning drainage had mainly net emissions during a 2-year monitoring period, whereas its well-drained counterpart showed net consumption of CH_4 (Regina et al. 2007).

Emissions from open ditches can have a large impact on the total greenhouse balance of a drained peatland (Schrier-Uijl et al. 2011; Minkkinen and Laine 2006). There are limited data on these emissions from Finnish croplands (Hyvonen et al. 2013). As it is known that the nutrient status of the drained area greatly affects the emission rate, it is likely that the open ditches around nutrient-rich cultivated fields are a high source of CH_4 emissions. However, most fields are subsurface drained, and the significance of these emissions is likely minor in the country scale.

2.3.2.3 Nitrous Oxide

In organic soil, peat decomposition is the main source of N_2O emissions and fertilization has minor importance. Mineralization of nitrogen from the peat can have the magnitude of several hundreds of kilograms per hectare annually, and this enables relatively high emission rates of N_2O regardless of fertilization (Leppelt et al. 2014). The annual emissions of N_2O have varied between 0.04 and 5.47 g m^{-2} in Finnish measurements (Table 2.3). The annual emissions are thus several folds compared to mineral soils.

The average annual emission rates have been higher for annual than perennial crops indicating tighter nitrogen cycle in the case of perennial grasses that are able to take up nutrients until late autumn and are tilled less frequently.

Like in mineral soils, the residual nitrogen after harvest is a substrate for N_2O production during the winter period, and about half of the annual emissions can occur between harvest and sowing. Climatic conditions in the winter have a large effect on the annual emissions. It was observed that a warm period in the winter that melted 10 cm of the frost induced a 100-fold increase in N_2O concentration of the soil profile (Regina et al. 2004). There was a clear difference to a similarly managed field in northern Finland where frost was constant, however, and concentrations of N_2O in the soil remained low for the whole winter and no production of N_2O was observed.

2.3.2.4 Climatic Impact

The net climatic impact of a hectare of cropland can be estimated by converting the emissions of CH_4 and N_2O to carbon dioxide (Myhre et al. 2013). The calculated values for net global warming potential of annual and perennial cropping are 32 and 18 t CO_2 eq. ha^{-1} per year suggesting almost double emission rates from annual compared to perennial crops (Table 2.3). However, a valid comparison requires comparing crop types within the same site. This data set is biased in the case of perennial crops as half of the observations come from high-yielding bioenergy crops. Even if this data set is too small for a robust comparison, the results point to the direction that less frequent soil disturbance slows down peat decomposition and thus diminishes the climatic impact of cultivation on organic soils.

2.4 Mitigation Options

Due to the short growing period prevailing in Finland, the most feasible management change to reduce the climatic impact of any type of agricultural soils would be reducing the period of bare fallow after harvest. This has not been studied in field experiments with greenhouse gas emission measurements, but the lower emissions rates from annual compared to perennial crops suggest that the longer the vegetated period annually, the smaller are the environmental effects of a cultivated field. Bare soil is prone to losses through runoff, leaching and gaseous emissions. A growing plant like a cover crop takes up nitrogen and has the potential to reduce leaching losses (Valkama et al. 2015) and potentially losses as N_2O if the risk of increased N_2O emission from the decomposing residues of the cover crop can be avoided. However, the risk of increasing the annual emissions with the presence of a cover crop is evident as revealed in the meta-analysis of Han et al. (2017) indicating that while the after-harvest emissions of N_2O are reduced with the presence of a cover crop, the annual emissions may not be. In any case, the extra crop residues of the cover crop have the potential for carbon stock increment in mineral soils (Poeplau and Don 2015). Including cover crops in rotation would compensate for the current low carbon input in crop residues and thus help to maintain carbon stocks and fertility of croplands. Other means of avoiding losses after harvest would be inclusion of autumn-sown crops in the rotations or spring tillage instead of autumn tillage. The above-mentioned practices may become more common as the climate warms and survival of the vegetation in the autumn period becomes more likely.

Also related to the northern location of Finland, there is the need to renew the grass swards every 3–4 years. If the renewal was done less frequently or selectively on only the poor areas of a field, this would likely reduce the losses related to tillage and bare soil. However, data on the effects of grass sward renewal on N_2O emissions is still scarce, and the results are too short term and show varying results (Buchen et al. 2017).

All measures improving the nitrogen use efficiency reduce available nitrogen for microbial processes and thus the emissions of N_2O. Management options related to this include optimizing fertilizer amount and timing as well as precision farming techniques that take the spatial differences within a field into account. However, the total use of nitrogen inputs in mineral fertilizers has already been reduced by 35% in 1990–2016, and thus further large-scale reductions may affect yields per hectare. This is not a desirable trend as it may lead to the need to clear more field area from native soils to maintain food production.

Reducing tillage intensity has been often found to increase the carbon stocks of mineral soils (Sainju 2016), and there is also some evidence of reduced carbon losses from no-till organic soils (Elder and Lal 2008; Regina and Alakukku 2010). However, in conditions of northern countries with relatively high soil moisture, the emissions of N_2O may increase in the denser topsoil negating part of the favourable effects of minimum tillage (Gregorich et al. 2005; Sheehy et al. 2013). In addition, it is also not evident that reducing tillage would always increase soil carbon stocks as the stock may reduce in the deeper layers resulting in no change for the whole soil profile (Powlson et al. 2014).

In a country like Finland with high coverage of peat soils, the most promising mitigation options are found in management of cultivated organic soils (Klove et al. 2017). With the high hectare-based emissions, expected mitigation effects are higher than in mineral soils. The measures for reducing the climatic impacts of food production on peat soils can be divided into those that reduce the cultivated area and those that reduce the hectare-based emission rates while the cultivation continues. It has been found in many studies that abandoned fields on organic soil can still be relatively high sources of greenhouse gases (Maljanen et al. 2012a, 2013). Thus, it would be beneficial to have well-planned strategies for the after-use of cultivated peat soils instead of uncontrolled abandoning.

Most realistic option for reducing the area of peat soils under cultivation in Finland is afforestation. However, in conventional afforestation management, drainage and evapotranspiration of the tree stand maintain the groundwater table low and peat decomposition continues in the layer above the groundwater table. Also, the emissions of N_2O may continue at a rate similar to cultivated soils (Maljanen et al. 2010a, 2012a). Afforestation could be a preferential strategy for fields that have been cultivated for several decades and already have lost most of the peat layer. Using tree species like alder or birch afforestation with high groundwater table level may also be feasible (Wichtmann et al. 2016).

As the peat has accumulated due to reduced decomposition of plant residues caused by a high groundwater level, evidently the most effective way to reduce the emissions of cultivated peat soils would be to raise the water table as high as possible. Total rewetting is possible for the fields that are not necessary for food production and in locations with abundant water reserves. The risk of high CH_4 emissions (Schafer et al. 2012) and nutrient losses to watercourses (Kieckbusch and Schrautzer 2007) should be taken into account when planning such measures. There are at least two examples of successful rewetting cases on former agricultural soils with the result of emission neutrality or net sink effect (Herbst et al. 2013; Schrier-Uijl et al. 2014), and this option should be studied also in the Nordic conditions.

Some crops resistant to wet soil conditions can be produced in paludiculture (Wichtmann et al. 2016). Paludiculture enables maintaining production while the groundwater level is raised close to the soil surface. However, many crops that are suitable for such production have very limited markets. In Finland, this could be a future option to reduce greenhouse gas emissions in regions where agriculture is an important livelihood, but limited areas of mineral soils are available for agricultural activities.

High groundwater table reduces the yields of most agricultural crops and hinders the use of heavy machinery. Temporary raise of groundwater table is possible using controlled drainage which allows for decreasing the water table level whenever there is a need to use machines on the field (Osterholm et al. 2015). This is a more feasible option to implement greenhouse gas mitigation by raised water table and may become more common in the near future.

2.5 Gaps in Knowledge

The data survey on greenhouse gas emission measurements done in Finland shows that the uncertainties are still high in all categories. An especially high need exists for measurement results ranging over the full year in the case of carbon budget of mineral soils. Almost lacking are measurements of all greenhouse gases from drainage ditches. Acid sulphate soils deserve more attention as a potentially large emission source. Most of the reported flux measurements were done on the most typical crops and practices: grass ley in crop rotation or spring barley with autumn tillage. Thus, there is lack of knowledge on the effects of, e.g. a prolonged vegetated period or spring tillage which could potentially reduce losses to both atmosphere and watercourses after harvest. Also, data on the effects of raised water table on cultivated organic soils and the guidelines on how to manage such fields are largely lacking.

Climate policy drives the development of agricultural policies including more measures affecting the climatic impacts of agricultural production. From the policy viewpoint, mitigation measures are useless if their effects cannot be shown in statistics. It will be necessary to develop methods for reporting the effects of the most prominent mitigation measures in the official emissions statistics. For this, future research should be designed to cover as many management practices as possible to build a large enough database for proving the effects of management changes. This is also important to convince farmers on the beneficial effects of their activities.

References

Bhattarai HR, Virkajarvi P, Yli-Pirilda P, Maljanen M (2018) Emissions of atmospherically important nitrous acid (HONO) gas from northern grassland soil increases in the presence of nitrite (NO_2^-). Agric Ecosyst Environ 256:194–199

Buchen C, Well R, Helfrich M, Fuss R, Kayser M, Gensior A, Benke M, Flessa H (2017) Soil mineral N dynamics and N_2O emissions following grassland renewal. Agric Ecosyst Environ 246:325–342

Denmead OT, Macdonald BCT, Bryant G, Naylor T, Wilson S, Griffith DWT, Wang WJ, Salter B, White I, Moody PW (2010) Emissions of methane and nitrous oxide from Australian sugarcane soils. Agric For Meteorol 150(6):748–756

Elder JW, Lal R (2008) Tillage effects on physical properties of agricultural organic soils of north central Ohio. Soil Tillage Res 98(2):208–210

Gregorich E, Rochette P, Vandenbygaart A, Angers D (2005) Greenhouse gas contributions of agricultural soils and potential mitigation practices in Eastern Canada. Soil Tillage Res 83 (1):53–72

Gronlund A, Hauge A, Hovde A, Rasse DP (2008) Carbon loss estimates from cultivated peat soils in Norway: a comparison of three methods. Nutr Cycl Agroecosyst 81(2):157–167

Guo L, Gifford R (2002) Soil carbon stocks and land use change: a meta analysis. Glob Change Biol 8(4):345–360

Han Z, Walter MT, Drinkwater LE (2017) N_2O emissions from grain cropping systems: a meta-analysis of the impacts of fertilizer-based and ecologically-based nutrient management strategies. Nutr Cycl Agroecosyst 107(3):335–355

Heikkinen J, Ketoja E, Nuutinen V, Regina K (2013) Declining trend of carbon in Finnish cropland soils in 1974–2009. Glob Change Biol 19(5):1456–1469

Herbst M, Friborg T, Schelde K, Jensen R, Ringgaard R, Vasquez V, Thomsen AG, Soegaard H (2013) Climate and site management as driving factors for the atmospheric greenhouse gas exchange of a restored wetland. Biogeosciences 10(1):39–52

Hyvonen NP, Huttunen JT, Shurpali NJ, Lind SE, Marushchak ME, Heitto L, Martikainen PJ (2013) The role of drainage ditches in greenhouse gas emissions and surface leaching losses from a cutaway peatland cultivated with a perennial bioenergy crop. Boreal Environ Res 18 (2):109–126

Karhu K, Wall A, Vanhala P, Liski J, Esala M, Regina K (2011) Effects of afforestation and deforestation on boreal soil carbon stocks—comparison of measured C stocks with Yasso07 model results. Geoderma 164(1–2):33–45

Katterer T, Bolinder MA, Andren O, Kirchmann H, Menichetti L (2011) Roots contribute more to refractory soil organic matter than above-ground crop residues, as revealed by a long-term field experiment. Agric Ecosyst Environ 141(1–2):184–192

Kieckbusch JJ, Schrautzer J (2007) Nitrogen and phosphorus dynamics of a re-wetted shallow-flooded peatland. Sci Total Environ 380(1–3):3–12

Klove B, Berglund K, Berglund O, Weldon S, Maljanen M (2017) Future options for cultivated Nordic peat soils: can land management and rewetting control greenhouse gas emissions? Environ Sci Policy 69:85–93

Koponen H, Martikainen P (2004) Soil water content and freezing temperature affect freeze-thaw related N_2O production in organic soil. Nutr Cycl Agroecosyst 69(3):213–219

Leppelt T, Dechow R, Gebbert S, Freibauer A, Lohila A, Augustin J, Droesler M, Fiedler S, Glatzel S, Hoeper H, Jaerveoja J, Laerke PE, Maljanen M, Mander U, Maekiranta P, Minkkinen K, Ojanen P, Regina K, Stromgren M (2014) Nitrous oxide emission budgets and land-use-driven hotspots for organic soils in Europe. Biogeosciences 11(23):6595–6612

Lind SE, Shurpali NJ, Peltola O, Mammarella I, Hyvonen N, Maljanen M, Raty M, Virkajarvi P, Martikainen PJ (2016) Carbon dioxide exchange of a perennial bioenergy crop cultivation on a mineral soil. Biogeosciences 13(4):1255–1268

Lohila A, Aurela M, Tuovinen J, Laurila T (2004) Annual CO_2 exchange of a peat field growing spring barley or perennial forage grass. J Geophys Res Atmospheres 109(D18):D18116

Luke (2016) Ruoka- ja luonnonvaratilastojen e-vuosikirja 2016. [In Finnish]. Available at: http://stat.luke.fi/ruoka-ja-luonnonvaratilastojen-e-vuosikirja-2016-2016_fi

Maljanen M, Martikainen P, Walden J, Silvola J (2001) CO_2 exchange in an organic field growing barley or grass in eastern Finland. Glob Change Biol 7(6):679–692

Maljanen M, Liikanen A, Silvola J, Martikainen P (2003a) Methane fluxes on agricultural and forested boreal organic soils. Soil Use Manag 19(1):73–79

Maljanen M, Liikanen A, Silvola J, Martikainen P (2003b) Nitrous oxide emissions from boreal organic soil under different land-use. Soil Biol Biochem 35(5):689–700

Maljanen M, Komulainen V, Hytonen J, Martikainen P, Laine J (2004) Carbon dioxide, nitrous oxide and methane dynamics in boreal organic agricultural soils with different soil characteristics. Soil Biol Biochem 36(11):1801–1808

Maljanen M, Kohonen A, Virkajarvi P, Martikainen PJ (2007a) Fluxes and production of N_2O, CO_2 and CH_4 in boreal agricultural soil during winter as affected by snow cover. Tellus Ser B Chem Phys Meteorol 59(5):853–859

Maljanen M, Martikkala M, Koponen HT, Virkajarvi P, Martikainen PJ (2007b) Fluxes of nitrous oxide and nitric oxide from experimental excreta patches in boreal agricultural soil. Soil Biol Biochem 39(4):914–920

Maljanen M, Virkajarvi P, Hytonen J, Oquist M, Sparrman T, Martikainen PJ (2009) Nitrous oxide production in boreal soils with variable organic matter content at low temperature—snow manipulation experiment. Biogeosciences 6(11):2461–2473

Maljanen M, Sigurdsson BD, Guomundsson J, Oskarsson H, Huttunen JT, Martikainen PJ (2010a) Greenhouse gas balances of managed peatlands in the Nordic countries—present knowledge and gaps. Biogeosciences 7(9):2711–2738

Maljanen M, Hytonen J, Martikainen PJ (2010b) Cold-season nitrous oxide dynamics in a drained boreal peatland differ depending on land-use practice. Can J For Res 40(3):565–572

Maljanen M, Shurpali N, Hytonen J, Makiranta P, Aro L, Potila H, Laine J, Li C, Martikainen PJ (2012a) Afforestation does not necessarily reduce nitrous oxide emissions from managed boreal peat soils. Biogeochemistry 108(1–3):199–218

Maljanen M, Virkajarvi P, Martikainen PJ (2012b) Dairy cow excreta patches change the boreal grass swards from sink to source of methane. Agric Food Sci 21(2):91–99

Maljanen M, Hytonen J, Makiranta P, Laine J, Minkkinen K, Martikainen PJ (2013) Atmospheric impact of abandoned boreal organic agricultural soils depends on hydrological conditions. Boreal Environ Res 18(3–4):250–268

Minkkinen K, Laine J (2006) Vegetation heterogeneity and ditches create spatial variability in methane fluxes from peatlands drained for forestry. Plant Soil 285(1–2):289–304

Muller C, Kammann C, Ottow J, Jager H (2003) Nitrous oxide emission from frozen grassland soil and during thawing periods. J Plant Nutr Soil Sci 166(1):46–53

Myhre G, Shindell D, Bréon F, Collins W, Fuglestvedt J, Huang J, Koch D, La-Marque J, Lee D, Mendoza B, Nakajima T, Robock A, Stephens G, Takemura T, Zhang H (eds) (2013) Anthropogenic and natural radiative forcing. Cambridge University Press, Cambridge, United Kingdom and New York, NY, USA

Osterholm P, Virtanen S, Rosendahl R, Uusi-Kamppa J, Ylivainio K, Yli-Halla M, Maensivu M, Turtola E (2015) Groundwater management of acid sulfate soils using controlled drainage, by-pass flow prevention, and subsurface irrigation on a boreal farmland. Acta Agric Scand Sect B Soil Plant Sci 65:110–120

Petersen S, Regina K, Pollinger A, Rigler E, Valli L, Yamulki S, Esala M, Fabbri C, Syvasalo E, Vinther F (2006) Nitrous oxide emissions from organic and conventional crop rotations in five European countries. Agric Ecosyst Environ 112(2–3):200–206

Petersen SO, Hoffmann CC, Schafer C, Blicher-Mathiesen G, Elsgaard L, Kristensen K, Larsen SE, Torp SB, Greve MH (2012) Annual emissions of CH_4 and N_2O, and ecosystem respiration, from eight organic soils in Western Denmark managed by agriculture. Biogeosciences 9(1):403–422

Poeplau C, Don A (2015) Carbon sequestration in agricultural soils via cultivation of cover crops—a meta-analysis. Agric Ecosyst Environ 200:33–41

Powlson DS, Stirling CM, Jat ML, Gerard BG, Palm CA, Sanchez PA, Cassman KG (2014) Limited potential of no-till agriculture for climate change mitigation. Nat Clim Change 4 (8):678–683

Regina K, Alakukku L (2010) Greenhouse gas fluxes in varying soils types under conventional and no-tillage practices. Soil Tillage Res 109(2):144–152

Regina K, Syvasalo E, Hannukkala A, Esala M (2004) Fluxes of N_2O from farmed peat soils in Finland. Eur J Soil Sci 55(3):591–599

Regina K, Virkajärvi P, Saarijärvi K, Maljanen M (2006) Kasvihuonekaasupäästöt laitumilta ja suojakaistoilta. In: Virkajärvi, P., Uusi-Kämppä, J. (Eds.), Laitumien ja suojavyöhykkeiden ravinnekierto ja ympäristökuormitus. Maa- ja elintarviketalous 76:88–100. [in Finnish, abstract in English]

Regina K, Pihlatie M, Esala M, Alakukku L (2007) Methane fluxes on boreal arable soils. Agric Ecosyst Environ 119(3–4):346–352

Regina K, Kaseva J, Esala M (2013) Emissions of nitrous oxide from boreal agricultural mineral soils-Statistical models based on measurements. Agric Ecosyst Environ 164:131–136

Sainju UM (2016) A global meta-analysis on the impact of management practices on net global warming potential and greenhouse gas intensity from cropland soils. PLoS ONE 11(2):e0148527

Schafer C, Elsgaard L, Hoffmann CC, Petersen SO (2012) Seasonal methane dynamics in three temperate grasslands on peat. Plant Soil 357(1–2):339–353

Schrier-Uijl AP, Veraart AJ, Leffelaar PA, Berendse F, Veenendaal EM (2011) Release of CO_2 and CH_4 from lakes and drainage ditches in temperate wetlands. Biogeochemistry 102(1–3): 265–279

Schrier-Uijl AP, Kroon PS, Hendriks DMD, Hensen A, van Huissteden J, Berendse F, Veenendaal EM (2014) Agricultural peatlands: towards a greenhouse gas sink—a synthesis of a Dutch landscape study. Biogeosciences 11(16):4559–4576

Sheehy J, Six J, Alakukku L, Regina K (2013) Fluxes of nitrous oxide in tilled and no-tilled boreal arable soils. Agric Ecosyst Environ 164:190–199

Shurpali NJ, Hyvonen NP, Huttunen JT, Clement RJ, Reichstein M, Nykanen H, Biasi C, Martikainen PJ (2009) Cultivation of a perennial grass for bioenergy on a boreal organic soil—carbon sink or source? Glob Change Biol Bioenergy 1(1):35–50

Simek M, Virtanen S, Kristufek V, Simojoki A, Yli-Halla M (2011) Evidence of rich microbial communities in the subsoil of a boreal acid sulphate soil conducive to greenhouse gas emissions. Agric Ecosyst Environ 140(1–2):113–122

Simek M, Virtanen S, Simojoki A, Chronakova A, Elhottova D, Krigtufek V, Yli-Halla M (2014) The microbial communities and potential greenhouse gas production in boreal acid sulphate, non-acid sulphate, and reedy sulphidic soils. Sci Total Environ 466:663–672

Singh P, Heikkinen J, Ketoja E, Nuutinen V, Palojarvi A, Sheehy J, Esala M, Mitra S, Alakukku L, Regina K (2015) Tillage and crop residue management methods had minor effects on the stock and stabilization of topsoil carbon in a 30-year field experiment. Sci Total Environ 518:337–344

Syvasalo E, Regina K, Pihlatie M, Esala M (2004) Emissions of nitrous oxide from boreal agricultural clay and loamy sand soils. Nutr Cycl Agroecosyst 69(2):155–165

Syvasalo E, Regina K, Turtola E, Lemola R, Esala M (2006) Fluxes of nitrous oxide and methane, and nitrogen leaching from organically and conventionally cultivated sandy soil in western Finland. Agric Ecosyst Environ 113(1–4):342–348

Teepe R, Vor A, Beese F, Ludwig B (2004) Emissions of N_2O from soils during cycles of freezing and thawing and the effects of soil water, texture and duration of freezing. Eur J Soil Sci 55 (2):357–365

UNFCCC (2018) Annual submissions of greenhouse gas inventories. Available at: https://unfccc. int/process/transparency-and-reporting/reporting-and-review-under-the-convention/greenhouse-gas-inventories-annex-i-parties/national-inventory-submissions-2018

Valkama E, Lemola R, Kankanen H, Turtola E (2015) Meta-analysis of the effects of undersown catch crops on nitrogen leaching loss and grain yields in the Nordic countries. Agric Ecosyst Environ 203:93–101

Virkajarvi P, Maljanen M, Saarijarvi K, Haapala J, Martikainen PJ (2010) N_2O emissions from boreal grass and grass-clover pasture soils. Agric Ecosyst Environ 137(1–2):59–67

Wichtmann W, Schröder C, Joosten H (eds) (2016) Paludiculture—productive use of wet peatlands—climate protection—biodiversity—regional economic benefits. Schweizerbart Science Publishers, Stuttgart, Germany

Yli-Halla M, Puustinen M, Koskiaho J (1999) Area of cultivated acid sulfate soils in Finland. Soil Use Manag 15(1):62–67

Chapter 3
Greenhouse Gas Exchange from Agriculture in Italy

Anna Dalla Marta and Leonardo Verdi

Abstract Together with industry, transport and energy sectors, agriculture is one of the main sources of greenhouse gas (GHG) emissions from human activities. In particular, intensive breeding, fertilization and fuel combustion for traction are the most impactful factors in terms of global GHGs. Italy is not an exception, and due to the high variability of environmental and morphological conditions of the country, a wide range of agricultural systems is adopted with different GHG emissions' potential. Italian agriculture accounts for 1.5% of total carbon dioxide (CO_2), 60.6% of total methane (CH_4), 68% of total nitrous oxide (N_2O) and 94% of total ammonia (NH_3) that is an indirect source of N_2O as well as the main source of N-volatilization from agriculture. CO_2 emissions are primarily produced from tractors on croplands, and based on technological level, there are no significant differences along the country. Because of the high concentration of paddy fields and intense breeding, the northern part of Italy is responsible for the majority of CH_4 and N_2O (and NH_3) production. Paddy lands are, in fact, the main sources of CH_4 emissions that are produced by flooded fields from anaerobic micro-organisms. In Italy, paddy lands cover more than 200,000 ha (51% of total EU paddy lands) following the course of Po River. However, N-based emissions are mainly produced by intense breeding, in particular through enteric fermentation and manure storage and spreading. Italian agriculture accounts for more than 138,000 cattle farms (6 million cattle) and 145,000 pig farms (8.7 million pigs) that are principally located in the central-northern part of the country. In addition, the intensive use of fertilizers contributes to N-based emissions through nitrification/denitrification and N-volatilization processes into the soil.

A. Dalla Marta (✉) · L. Verdi
Department of Agrifood Productions and Environmental Sciences,
University of Florence, Piazzale delle Cascine 18 – 50144, Florence, Italy
e-mail: anna.dallamarta@unifi.it

© Springer Nature Singapore Pte Ltd. 2019
N. Shurpali et al. (eds.), *Greenhouse Gas Emissions*, Energy, Environment,
and Sustainability, https://doi.org/10.1007/978-981-13-3272-2_3

3.1 Introduction

In Italy, the monitoring of GHG emissions is implemented by Istituto Superiore per la Protezione e la Ricerca Ambientale (ISPRA) for which the inventory of GHG emissions, an official tool for the verification of international commitments on the protection of the atmospheric environment with the United Nations Framework Convention on Climate Change (UNFCCC) and the Kyoto Protocol, is an important institutional commitment. In the present section, we reported a review of GHG emissions produced by Italian agriculture based on the most recent census of the Ministry of Agriculture, the last report on the state of the environment (ISPRA) and on the most recent national literature.

In Italy, emissions of GHG are estimated through the adoption of specific emission factors for each emission source. Estimates of GHG emissions from agriculture and Land Use, Land Use Change and Forestry (LULUCF) are implemented according to the IPCC methodology. Based on the requirements of the reference methodology, the inventory of GHG emissions produced by agriculture includes the estimation of two greenhouse gases, methane (CH_4) and nitrous oxide (N_2O). The emission sources for which emissions are estimated are enteric fermentation (CH_4 emissions), the management of manure and slurries (CH_4 and N_2O), agricultural soils (N_2O), the paddy fields (CH_4) and combustion of agricultural residues (CH_4 and N_2O).

Due to the high variability of environments and climate within the country, several agricultural systems are adopted. Italy ranges from typical Mediterranean agroecosystems where drought and high-temperature-resistant crops as citrus and olive are cultivated, to alpine climate where only mountain farming is possible.

In general, the highest amount of CO_2 eq derives from husbandry-related activities (enteric fermentation, manure management/application) and mainly involves CH_4 (Fig. 3.1). Intense livestock systems represent a relevant factor of Italian agriculture with more than 138,000 cattle farms (6 million cattle), 145,000 pig farms (8.7 million pigs), 76,000 sheep farms (6.7 million of sheep) and 80,000 chicken farms (over 157 million of chickens) that are principally located in the central-northern part of the country. In particular, 74% of emissions from livestock are produced by cattle; it is estimated that one cow emits around 2.6 tons of carbon dioxide equivalent (CO_2 eq) per year (compared to 0.9 and 0.1 tons of CO_2 eq from one sheep and one pig per year, respectively) (ISPRA 2017).

Further, Italian agriculture accounts for more than 200,000 ha of paddy lands (51% of total EU paddy lands) dislocated along the course of Po River.

Based on collected data, in 2015 Italian agriculture was responsible of 6.9% of global GHG emissions and represented the third emission source after energy sector and industry production process. In particular, enteric fermentation represented 46.0% of Italian GHG emissions followed by croplands (29.9%), manure and slurries management (17.0%) and paddy lands (5.6%).

Nevertheless, GHG emissions from Italian agriculture in 2015 were 30 Mt CO_2 eq, that is 15.9% less compared to 1990 when emissions were 35.6 Mt CO_2 eq

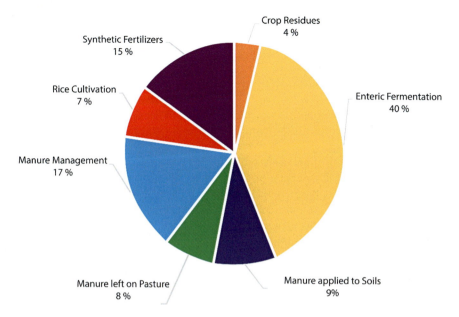

Fig. 3.1 Emission of CO$_2$ eq (average 1990–2016) from different agricultural activities in Italy (adapted from FAOSTAT)

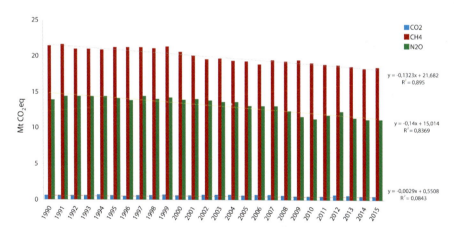

Fig. 3.2 Trend of GHG emissions from Italian agriculture from 1990 to 2015 (*Source of data* ISPRA)

(Fig. 3.2). This trend was mainly related to the reduction in the number of livestock (−30% of cattle and −22% of sheep), the reduction of the use of N-based synthetic fertilizers and the contraction of cultivated land. In particular, it was observed that from 1990 to 2015, croplands in Italy decreased by 17%. In this sense, policies promoting high-efficiency agricultural systems, adoption of modern technologies and renewable energy sources had a great impact. In addition, specific founds were dedicated to those agricultural management strategies that aim to reduce environmental pressure of agriculture, increase biodiversity and reduce climate change.

Following IPCC guidelines, by 2030 global agricultural emissions must decrease by 15% compared to present. IPCC also affirms that current mitigation strategies may produce a benefit not exceeding 40% of the set target. So, FAO advises further investments from governments on the definition of modern and more efficient agricultural strategies (precision farming, circular economy, conservation agriculture, etc.). A challenging topic is the maintenance of agricultural productivity to satisfy the increasing food demand with increasing population growth, yet achieving the intended reduction of GHG emissions and conservation of natural resources. In this sense, Italian agriculture, to maintain current food self-sufficiency (80% of national request), must achieve the reduction of GHG emissions without any further reduction in the area under crops and livestock density.

Italian agriculture, as agriculture at the global scale, contributes fairly modestly to CO_2 production. Main sources are represented by soil organic matter oxidation, crop and crop residues combustion and fuels for traction. Nevertheless, these emissions are almost completely balanced by biological uptake, and compared to other gases such as CH_4, CO_2 represents a minor issue for global warming. On the other hand, NH_3, not a GHG but a precursor to N_2O production, represents a relevant loss of N into the atmosphere.

The amount of emissions of the three mentioned gases, CH_4, N_2O and NH_3, varies across the country (Fig. 3.3).

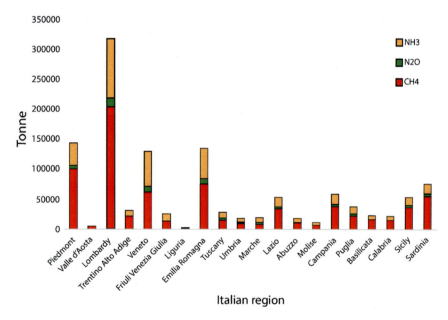

Fig. 3.3 Emissions (Mt) of CH_4, N_2O and NH_3 from agriculture in the different Italian regions (*Source of data* ISTAT 2005)

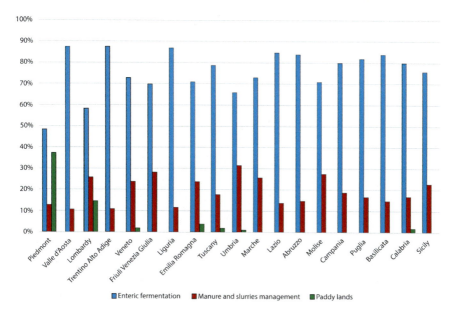

Fig. 3.4 CH$_4$ emissions in each Italian region from different sources (*Source* Adapted from ISPRA 2017)

3.2 Carbon Emissions Produced by Italian Agriculture

Beside CO$_2$, CH$_4$ is the main carbon emission produced by Italian agriculture. One of the most impactful agricultural activities in terms of CH$_4$ is represented by intense livestock systems that produce CH$_4$ through the enteric fermentation of animals (70% of total CH$_4$ emissions from agriculture) and the management of manure and slurries (20% of total CH$_4$ emissions from agriculture). Another important source of CH$_4$ (10% of total CH$_4$ emissions) is rice cultivation; paddy lands are mainly located in the north of the country in Piedmont (37%), Lombardy (14%), Veneto and Emilia-Romagna regions. Thus, Italian agriculture is responsible of 15.3% of total CH$_4$ emissions produced in the country (Fig. 3.4).

3.3 Nitrogen Emissions Produced by Italian Agriculture

In 2015, N emissions from Italian agriculture accounted for 342.2 Kt with a net reduction of 17.7% compared to 1990 (ISPRA 2017). This reduction is again mainly related to the decrease in the number of livestock, the reduction of croplands, the reduction of the use of N-based fertilizers and the adoption of high-efficiency fertilization strategies. Nevertheless, ammonia (NH$_3$) is the main

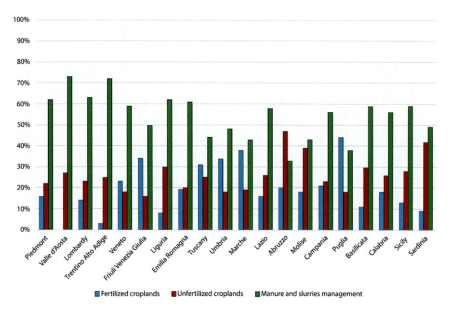

Fig. 3.5 N$_2$O emissions in each Italian region from different sources (*Source* Adapted from ISPRA 2017)

source of N losses into the atmosphere representing the 90.9% of N emissions, followed by N$_2$O that is the 6.9% (Fig. 3.5).

NH$_3$ is mainly produced in the north of Italy, Sardinia and north of Puglia, where intense livestock is more abundant and where the adoption of NH$_4$-based fertilizer is still relevant. Again, N$_2$O production is mainly concentrated in the north of Italy where, in addition to the intense adoption of N-based fertilizers, annual precipitation and soil water content are higher compared to the rest of the country. This is also true for the regions in the south of Italy where annual precipitation is lower but the adoption of irrigation is relevant (Fig. 3.6).

3.4 Mitigation Strategies

Due to the high number of intense livestock systems, Italian agriculture in the last few years is showing an intense increase in the number of biogas plants. In particular, the number of plants has grown from 10 to nearly 900 (Fabbri et al. 2013) in a few years, and many more plants are under construction. According to a recent census (Fabbri et al. 2013), at present there are more than 1000 biogas-operating plants (Carrosio 2013) making Italy the second larger biogas producer in Europe after Germany. Biogas production is an excellent way of using organic waste for energy generation, followed by the recycling of the digested substrate (digestate) as fertilizer (Comparetti et al. 2013; Maucieri et al. 2016) with low emissions'

3 Greenhouse Gas Exchange from Agriculture in Italy

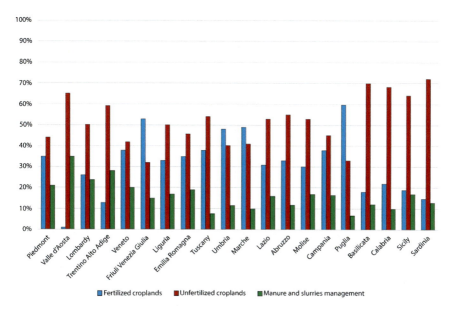

Fig. 3.6 NH$_3$ emissions in each Italian region from different sources (*Source* Adapted from ISPRA 2017)

potential. In fact, digestate use as replacing fertilizer represents an interesting strategy to reduce GHG emissions from agriculture while maintaining satisfying crop yield levels. Digestate is composed of two phases: a liquid fraction, rich in water and N-easy available compounds for crops, and a solid fraction rich in organic matter. The liquid fraction is the more interesting for fertilization purposes, but special attention should be paid to its management and spreading in order to achieve the highest possible agronomic efficiency while minimizing emissions.

For each Italian region, as well as in the rest of Europe, specific regulations provide guidance on the appropriate techniques for digestate management, and for slurries in general, for the reduction on N losses into the atmosphere. For instance, Tuscany region adopted the Regulation 46R/2008 for the protection of water from pollution by defining the allowed management techniques. The most adopted strategy is the use of different kind of coverages (floating polyethylene systems, clay balls, etc.) for digestate storage lagoons. Dinuccio and Balsari (2011) observed that covering digestate lagoons may reduce NH$_3$ emissions from 76% up to 99% based on the type of covering techniques adopted.

On the other hand, regional regulations provide information about digestate spreading techniques. In particular, incorporation of digestate into the soil is a mandatory strategy to reduce GHG emissions. Due to its high N–NH$_4^+$ content, when digestate is spread on soil surface a relevant volatilization of NH$_3$-based compounds occurs, with consequent atmospheric pollution and N losses. Moreover, the consequent soil enrichment in organic C and organic N compounds favours micro-organism's activities with a consequent increase of respiration and CO_2 emissions. Thus, incorporation of digestate into the soil instead of surface spreading strongly reduces GHG emissions' risks representing an efficient mitigation strategy.

Concerning paddy lands, a recent study (Lagomarsino et al. 2016) observed that the alternation of wetting and drying is an efficient strategy to reduce CH_4 emissions.

3.5 The Regulation Framework

In Italy, a "National System for the implementation of the National Inventory of Greenhouse Gases" (National System) was established, according to the obligations related to the implementation of the Kyoto Protocol, for the monitoring and accounting of GHGs. The Italian National System was established in 2008 and designated ISPRA as the responsible Institution for the National Inventory of GHGs, as well as the collection of basic data and the implementation of a program to check and guarantee their quality. The national register of agroforestry carbon reservoirs, which is an integral part of the National System, is the tool for the certification of GHGs fluxes deriving from afforestation, reforestation, deforestation and forest management activities, and was established in 2008 at the Italian Ministry for the Environment and Protection of the Territory and the Sea.

Beside the monitoring and accounting by the National System, the policies for the mitigation of emissions in the agricultural sector are part of the European and international framework of climate action.

A legislative proposal presented by the Commission in 2016 (Effort Sharing Regulation) has set national targets for reducing emissions of polluting gases to maintain the commitments of the Paris agreements. In October 2014, EU adopted the 2030 framework for climate and energy. The framework includes the binding target of reducing emissions in the EU by at least 40% by 2030 compared to 1990 levels. In particular, the reduction of emissions in sectors, such as transport, agriculture, buildings and waste, the so-called non-Emissions Trading System (non-ETS) sectors, must be 30% compared to 2005. To ensure that all countries participate in such reduction, the Effort Sharing Regulation has set country-by-country targets. The reduction target for Italy is −33% by 2030, compared to 2005.

Among the financing methods to favour the transition to a low-carbon economy of agriculture, the Common Agricultural Policy (CAP) plays a central role, by establishing an appropriate system of incentives. With the new CAP in fact, the focus on environmental sustainability is guaranteed by the introduction of the "greening", under which 30% of the national budget available for direct payments to farmers should be subject to the compliance with sustainable agricultural practices. Further, another positive boost comes from the implementation of Rural Development Plans (RDPs) called to address the four challenges of the "Health

Check" of CAP which are climate change, renewable energy, water resources management and biodiversity. The majority of RDPs supported measures for the reduction of GHGs and recognized the reduction of N surplus as the main way to fulfil the target. The European Agricultural Fund for Rural Development (EAFRD) includes several measures to encourage investment to improve the performance and sustainability of farms. In particular, with regard to climate action, specific agri-environment–climate measures are designed to encourage farmers to protect and enhance the environment on their farmland by paying them for the provision of environmental services. The main purpose of agri-environment–climate payments is "the introduction or maintenance of agricultural practices that contribute to mitigating climate change or that promote adaptation to them and that are compatible with the protection and improvement of the environment, landscape, natural resources, soil and genetic diversity", and it is the only mandatory rural development measure for Member States (but voluntary for farmers), Italy included.

The main environmental regulations impacting on the GHG emissions are Nitrates Directive, National Emissions Ceiling Directive (NEC), the Integrated Pollution Prevention and Control Directive (IPPC) and the Water Framework Directive (WFD). Concerning the Nitrates Directive, Italy made many progresses in the last years through the increase of the vulnerable areas (30% of UUA) and the introduction of a Best Agricultural Practices Code adopted by farmers interested by the regulation. The NEC Directive establishes maximum thresholds for each Member State for the main pollutants responsible for acidification, eutrophication and ozone-related pollution. In this sense, NH_3 is the most important pollutant deriving from agriculture. Italy has complied with the national emission limit for NH_3 set for the year 2010 at 419 kt (thousands of tons). The achievement of the objective was mainly due to the emissions trend of the agricultural sector and to the introduction of appropriate technologies due to the IPPC Directive (ISPRA 2017). Although not directly related to GHGs reduction, this addresses the implementation of the best available technologies (BAT) for the control of industrial pollution. For agriculture, the IPPC involves intensive breeding and mainly focuses on NH_3 emissions abatement so that it also impacts on gas emissions.

The revision of the NEC Directive (2016/2284) established the new reduction targets for 2020 and 2030. In particular, for Italy these targets are equal to 400.61 kt of national NH_3 emissions in 2020 (−5% compared to 2005) and 354.22 kt of national NH_3 emissions in 2030 (−16% compared to 2005) (ISPRA 2017).

3.6 Conclusions

Boosted by the new CAP and national guidelines, an intense adoption of sustainable agricultural management strategies is leading to a reduction of the emissions produced by Italian agriculture (ISPRA 2017). The reduction of the environmental impact is mainly related to the adoption of those strategies aiming at reducing agricultural inputs (fertilizers, pesticides, herbicides, etc.) and, consequentially,

GHG emissions. In particular, the reduction in the use of mineral fertilizers and the increasing of low-input and organic farming are driving Italian agriculture towards a more sustainable scenario. Moreover, the relevant amount of natural habitat in Italy is mainly driven by the adoption of extensive and conservative agricultural practices. In this sense, Italy represents one of the leaders in Europe for both the number and extension of such areas (ISPRA 2017).

In accordance with the last census (2017), Italian agriculture complied with the limits defined by EU guidelines. In particular, a significant reduction of N emissions was observed. Nevertheless, agriculture still represents the main source of national NH_3 emissions (93%) and the fifth source of GHG emissions (ISPRA 2017). In this regard, special attention is needed to maintain emissions at a sustainable level. However, due to the high variability of natural habitats and agricultural systems, Italy has a variable GHG emissions' potential. This implies the necessity of a dense monitoring network, which is still missing in Italy. In particular, it was affirmed by Minoli et al. (2016) that for NH_3 "a new series of measurements would be strongly needed". Nowadays, the most widely used strategy to investigate and define the trend of GHG emissions from Italian agriculture is represented by model simulations and by the adoption of specific emission factors. However, an improvement of actual databases of direct GHG emissions measurements is strongly needed to obtain reliable information and to propose effective mitigation strategies.

References

Carrosio G (2013) Energy production from biogas in the Italian countryside: policies and organizational models. Energ Policy 63:3–9. https://doi.org/10.1016/j.enpol.2013.08.072

Comparetti A, Febo P, Greco C, Orlando S (2013) Current state and future of biogas and digestate produciton. Bulg J Agric Sci 19:1–14

Dinuccio E, Balsari P (2011) Alcune soluzioni per la riduzione delle emisisoni di ammoniaca dallo stoccagio dei liquami zootecnici. Convegno di Medio Termine dell'Associazione Italiana di Ingegneria Agraria, Belgirate, 22–24 settembre 2011

Fabbri C, Labartino N, Manfredi S, Piccinini S (2013) Biogas, il settore è strutturato e continua a crescere. L'Informatore Agrario 11:11–18

ISPRA (2017) Annuario dei dati ambientali 2017. ISBN 978-88-448-0863-1

Lagomarsino A, Agnelli AE, Linquist B, Adviento-Borbe MA, Agnelli A, Gavina G, Ravaglia S, Ferrara RM (2016) Alternate wetting and drying of rice reduced CH4 emissions but triggered N_2O peaks in a clayey soil of central Italy. Pedosphere 26(4):533–548

Maucieri C, Barbera AC, Borin M (2016) Effect of injection depth of digestate liquid fraction on soil carbon dioxide emission and maize biomass production. Ital J Agronom 11:6–11

Minoli S, Acutis M, Carozzi M (2016) NH3 emissions from land application of manures and N-fertilisers: a review of the Italian literature. Ital J Agrometeorol 3:5–24

http://dati.istat.it/Index.aspx?DataSetCode=DCSP_ALLEV

Chapter 4
GHG Emissions and Mitigation in Romanian Vineyards

Eleonora Nistor, Alina Georgeta Dobrei, Alin Dobrei and Narasinha Shurpali

Abstract In viticulture, the water requirement is low and organic fertilizers such as manure or organic matter from cover crops or compost made from pomace and lees are generally applied. Therefore, some specialists are of the opinion that viticulture is less polluting than other farm sectors. Nevertheless, measures for mitigating GHG emissions from vineyards and associated wine industries need to be adopted to preserve the quality of grapevine by-products. In viticulture, GHG emission mitigation can be achieved through appropriate methods of tillage, fertilization, harvesting, irrigation, vineyard maintenance, transport or wine marketing, etc. Besides CO_2, nitrous oxide (N_2O) and methane (CH_4) are produced from fertilizers and waste/wastewater management, respectively. As main GHG in vineyards, N_2O can have the same harmful impact as CO_2. Carbon is found in grape leaves, shoots, and even fruit pulp, roots, canes, trunk or soil organic matter. C sequestration in soil by using less tillage and tractor passing is one of the most efficient methods to reduce GHG in vineyards. However, many years are needed for detecting the resulting SOC changes. In the last decades, among other methods, cover crops have been used successfully for GHG mitigation and increasing soil fertility in vineyards. There is still limited information on practical methods in reducing emissions of greenhouse gases in viticulture. Therefore, this paper serves as an information base for researchers and industries working in the viti- and vinicultural sectors by providing knowledge concerning GHG dynamics under standard management approaches and principles. This helps ensure businesses are equipped with new information useful to build an efficient strategy to handle and mitigate GHG emissions.

E. Nistor (✉) · A. G. Dobrei · A. Dobrei
Banat University of Agricultural Sciences and Veterinary Medicine "King Michael I of Romania", Calea Aradului 119, Timişoara, Romania
e-mail: nisnoranisnora@gmail.com

N. Shurpali
Department of Environmental and Biological Sciences, Yliopistoranta 1 DE, PO Box 1627, 70211 Kuopio, Finland

© Springer Nature Singapore Pte Ltd. 2019
N. Shurpali et al. (eds.), *Greenhouse Gas Emissions*, Energy, Environment, and Sustainability, https://doi.org/10.1007/978-981-13-3272-2_4

4.1 Introduction

Global warming is a direct or indirect result of human activities (burning fossil fuels, changing land use, energy consumption, etc.) with an impact on the atmospheric composition (Carlton et al. 2015). At the 19th International Wine and Spirit Exhibition "Vinexpo Bordeaux", organized during 18–21 June 2017, viticulturists addressed different possibilities to save their vineyards and made a call for general mobilization to fight against global warming. Climate change will definitely affect the vine-growing boundaries. High temperatures in the wine-growing area will increase sugar and alcohol rate and decrease acidity, which affect the wine flavour and taste. The result will be a reduction in vineyards in warm areas, while new areas will appear in places that were previously too cold for the vine growing. Fires, late frosts, prolonged drought, diseases and pests, hail and storms are events that will worsen in the future, as warned by John P. Holdren (Harvard Physicist, International Expert in Climate Change and Energy—Holdren 2018).

For greenhouse gas emissions' mitigation, viticulturists and winemakers try new working methods for their own and global benefit. Some successful practices have been proven to be: growing of drought-resistant rootstocks varieties, cover crops, grazing middle rows or straw mulch, expanding vineyards in colder climates, reducing the amount or total abandonment of the use of pesticides and herbicides, improving fertilizer application and irrigation, using green fertilizers, nitrification inhibitors and wastewater treatment, biochar incorporation in soil (not only reduces GHG emissions but also contributes to increased productivity and soil quality). Undoubtedly, interactions of the environment, climate and soil conditions with grape variety needs, floor management are the agricultural practices that can control N_2O and CO_2 release from vineyards.

4.2 Viticulture in Romania

GHG emissions from Romanian agriculture (the equivalent of Mt CO_2 to 1,000 € gross added value in agriculture) are among the lowest in the EU 28. Within the EU 28, Romania ranks fifth among the lowest share of greenhouse gas emissions from agricultural production and for the main components—nitrous oxide (N_2O), carbon dioxide (CO_2) and methane (CH_4). Therefore, in many small family farms without financial possibility to buy machinery and chemical fertilizers, GHG emissions are low. The total concentration of all GHG emissions had reached 441 ppm CO_2 equivalent in 2014, which is an increase of about 3 ppm compared with 2013 and 34 ppm compared with data from 2004. Experts believe that even small increases in global warming will reduce crop yields and will increase yield variability in low latitude regions (Scrucca et al. 2018).

4.2.1 Romanian Vineyards

The soils and the climate in Romanian vineyards are diverse. Therefore, different varieties can be grown for table wines or for high-quality wines. Vineyards are planted from 25 m altitudes (Dobrogea area—Black Sea), up to 600–700 m in piedmont areas in Transylvanian Hillsides. Soil type varies from sandy or light soils to clayey or limestone (Toti et al. 2015).

During 2000–2016, world area under vineyards decreased from 7.8 to 7.5 Mha in 2016. Five countries hold 50% of the world vineyards; Spain ranks the first with 14%, followed by China (11%), France (10%), Italy (9%) and Turkey (75); Romania ranks 10th with 191 Kha in 2016. Currently, it has an area of 243,000 ha of vineyards (242,000 ha older and 1,000 ha newly planted vineyards). The wine grapes represent 82% of the total vineyards area with wine production reaching 5–6 million hl/per year (Tamas 2017).

In Romania, grapevine growing is an ancient tradition. During the Roman Empire with the conquests in Dacia (the present territory of Romania), it is certainly known that the grapevine was cultivated in large areas. Romans brought new grape varieties, new winemaking and pruning methods. Even in the years with poor wine production or when the vineyards were affected by the invasion of phylloxera, varieties from the Drăgăşani, Odobeşti, Cotnari or Tarnavele vineyards and the Romanian wines such as Grasa de Cotnari, Tămâioasa, Busuioaca or Black/White Feteasca were appreciated in international markets (Bărbulescu 2017).

Vineyards in Romania (about 37) are grouped into eight viticultural regions, the most extensive of which is Moldavian Hills, which covers almost 70,000 ha. Relief differences (altitude, slope and sun exposure), soil and climate influence the ripening period from one region to another for the same variety. A variety grown in eastern Romania matures earlier by about 1 month than in north-west of the country, and therefore, different wine grape varietals cover each region (Irimia et al. 2017) (Fig. 4.1).

In 2016, vineyard area in Romania was 258 860.83 ha (1.79% of agricultural land), with 8432.39 ha (3.26%) on soil of the first-class quality, 65016.23 ha (25.12%) on soil of the second-class quality, 80346.63 ha (31.04%) on soil of the third-class quality, 79242.80 ha (30.61%) on soils of the fourth-class quality and 25822.78 ha (9.98%) on soil of the fifth-class quality.

Transylvania Plateau vine-growing region (I) includes five vineyards (Tarnave, Alba, Sebes-Apold, Aiud and Lechinta) with 17 vineyards. Main production is white wine (Protected Denomination of Origin (PDO) and Protected Geographical Indication (PGI)), semi-sweet, sweet and sparkling wines.

Moldavia Hills vine-growing region (II) is the largest and includes 12 vineyards (Cotnari, Odobesti, Panciu, Dealul Bujorului, Iaşi, Coteşti, Huşi, Covurlui, Colinele Tutovei, Iveşti, Nicoreşti, Zeletin). Most wines are white (PDO or PGI) and sweet. Cotnari wines are included in the catalogue of the best wines in the world. Dry wine can be found in Odobesti, Panciu and Cotesti vineyards. Red wines are produced in small quantities.

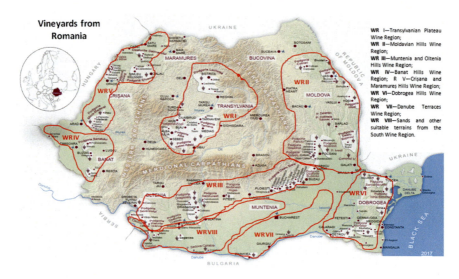

Fig. 4.1 Romanian vineyards and wine regions (WRs)

Oltenia and Muntenia hills' vine-growing region (III) includes eight vineyards (Dealu Mare, Sâmbureşti, Dealurile Buzăului, Ştefaneşti, Dealurile Craiovei, Plaiurile Drancei, Drăgăşani and Severin). In Samburesti are produced mainly red wines, while in the other vineyards, white and red wines labelled PDO or PGI from several varieties are produced.

In Banat Hills' vine-growing region (IV), and those two main vineyards (Recas, Buzias-Silagiu), are cultivated not only wine varieties but also table grapes (such as Chasselas, Black Hamburg, Muscat d'Adda or Victoria). White and red wines are labelled PDO (49%) or PGI (3%).

Crişana and Maramureş hills' vine-growing region (V) cultivates both white and red grapevine varieties, in Minis-Maderat, Valea lui Mihai, Diosig and Silvania vineyards for PDO (10%) and PGI (0.5%) wines and small quantities of sparkling wines.

Dobrogea hills' wine-growing region (VI) is located in the East of Romania near Black Sea and is well known from ancient times. Murfatlar, Istria-Babadag and Sarică-Niculiţel vineyards have around 17 342.70 ha, cultivated with table grapes and red/white wine varieties labelled as PDO (51%) and PGI (15%).

Danube Terrace wine growing (VII), can be found along the Danube on sandy soils and includes two main vineyards (Greaca and Ostrov—11 305.34 ha). The main production is table grapes and white wine varieties (PGI 3%).

Sands and other lands from South favourable wine region (VIII) include three vineyards (Calafat, Dacilor, Sadova-Corabia) on 13 029.40 ha cultivated with wine grape varieties labelled PGI (4%) and on small area with table grapes.

In Romania, vineyards cultivated with DOC wine varieties represents 15.1% from total vineyards area; PGI wine varieties from all vineyards hold 84.9%.

In 2015, Romania ranks first in the European Union by the number of vineyard owners (855.000 or 36% from total UE), but Romanian's owners hold the lowest average vineyards area (0.2 ha compared with French owners with 10.5 ha, or Austria—3.2 ha). The share of the vineyards area for table wines was 72.1% in Romania, followed by Bulgaria (38.4%) and Italy (26.2%).

Regardless of the global region that cultivates grapevine, global warming affects the growing area. As in other regions of the world, it is expected that in Romania and South-East Europe as well, climate change will have a major impact on the wine industry (Irimia et al. 2017). In order to cope with these changes, it is necessary to adopt new, adequate technologies that will contribute to the greenhouse gas emissions (GHG) mitigation. Viticulture, although less polluting than other agricultural sectors, has its contribution through fossil fuel consumption for maintenance or transport and energy related mainly to winemaking (Goode and Harrop 2011).

4.3 Greenhouse Gas Emissions in Viticulture

Grapevine, as a perennial plant with large canopy, is able to sequester much more CO_2 than annual crops. Unlike other industries, viticulture is not as polluting, but it is quite difficult to assess the level of greenhouse gas emissions (GHG), taking into account CO_2 emissions from the various management methods and technologies used in winemaking to the transport and distribution to the consumers (Brunori et al. 2016).

Organic soil matter from vineyards includes essential nutrients for plant nutrition and soil health, including carbon and nitrogen which are incorporated into the plant roots, microorganisms, dead tissue from plants or animals. Organic matter plays a major role in ensuring the physical, chemical and biological properties of the soil. The amount of carbon and nitrous oxide emissions from the soil is influenced by soil type, management, temperature, rainfall, vegetation (Suddick et al. 2010). Soil texture influences the carbon cycle; clay-rich soil retains organic matter between the particles and is hardly accessible to microorganisms (Krull et al. 2001).

Vineyard floor management influences the carbon and nitrous oxide loss from the soil and has major contribution in organic matter decomposition. In vineyards, greenhouse gas emissions (GHG—N_2O, CO_2, CH_4) result directly at the farm scale through soil tillage, indirect due to inputs (machines, seeds, fertilizers, pesticides, irrigation), or from grape juice fermentation, electrical power, gas and fuel consumption throughout the year, bottling and transport of wine to the consumers (Colman and Päster 2007).

Manure and compost, chopped and buried pruning debris, improve the carbon level in the soil. However, fertilizers applied in excess, on wrong place or very wet periods, lead to high GHG emissions (Toscano et al. 2013). Less herbicide applied in the vineyard increases plant biodiversity, the amount of carbon in soil and less CO_2 release to the atmosphere (Ball et al. 2014).

Vineyard irrigation contributes to N_2O and CO_2 emissions. Nitrous oxide (N_2O) is 300 times more dangerous for global warming than CO_2. High moisture content in the soil is equal with more N_2O emissions which are generated by microorganisms and organic matter decomposing. However, enough water in soil stimulates canopy growth and more carbon sequestration in plants (Robertson 1993).

4.3.1 Vineyard Carbon Dioxide Emissions and Potential Carbon Sequestration

Carbon atom is found in all organic plant or animals. Unfortunately, according to the latest statistics, Earth has passed the threshold of 400 ppm (parts per million) of carbon dioxide in the atmosphere, and there is very little chance that this limit will ever be lowered. By photosynthesis, plants convert CO_2 and H_2O into oxygen and carbohydrates. At night, photosynthesis stops, but the vine continues to respire. However, the amount of CO_2 released at night is lower compared to O_2 released or CO_2 sequestered over a day (Fraga et al. 2012). Simultaneously with surface photosynthesis, organic matter is decomposed in the soil; organic exudates from the roots, crops debris or the fall leaves are decomposed by the microorganisms and contribute to the soil fertility. Soils in general are a huge organic carbon pool (SOC), which is estimated to be 1 500–2 000 Pg C (1 Pg = 1015 g) till a depth of 1 m and at 2450 Pg till 2 m deep into the soil (about 2/3 from terrestrial carbon). In the same soil layers, inorganic carbon is stored up to 750 Pg (Carlier et al. 2009). It has been estimated that soil can be seized up to 20 Pg in 25 years (Zomer et al. 2017). The carbon stock in the vineyards remains constant without soil and other inputs in the soil (FAO 2017).

According to recent studies concerning winemaking and its life-cycle assessment (LCA) from the Oregon Region, viticultural practices contribute about 24%, winemaking with about 11% and packaging with 23% to the carbon footprint in wine life cycle. Distribution and transport rate of carbon footprint are around 13% but are greatly influenced by distances, type and models of bottling and packaging. Storage, consumption and refrigeration have 18% contribution to carbon footprint (CF), while disposal of wastes and packaging contributes with about 11% (Bonamente et al. 2016; Iannone et al. 2016; Benedetto 2013).

4.3.1.1 Direct and Indirect GHG Emissions

In vineyards and winemaking, there are the activities that result in greenhouse gases emissions and CO_2 sequestration. According to the Kyoto Protocol, OIV covers four greenhouse gases (CO_2, CH_4, N_2O, SF_6) and other two groups of hydrofluorocarbons

(HFCs) and perfluorocarbons (PFCs). Direct gas emission starts with the change of land use from previously forest or pasture ecosystems in vineyards (Suddick et al. 2010). By turning grassland into arable soil, around half of the carbon sink is lost in the early years (Johnston et al. 2017). The reverse process of C sequestration lasts up to 50 years for permanent grassland to accumulate the lost carbon stock. Researches by Garwood et al. (1977) have shown that grassland contains in the first 10 cm of soil a double amount of C compared to tillage soil from vineyards.

Vineyards "produce" CO_2 through vine respiration, soil tillage and fossil fuel, but "consume" CO_2 by photosynthesis. Nearly insignificant in vineyards, methane is released from anaerobic degradation process of organic matter, while nitrous oxide (N_2O) results from nitrogen fertilizers and transformations in the soil (Fraga et al. 2012). In the winemaking sector, refrigerant fluids release gases like hydrofluorocarbons (HFC), PFCs and SF_6; cold extraction and maceration, grape juice refrigeration, debourbage in white wines, pellicular maceration in red winemaking, controlled fermentations, cold storage of finished wines, amicrobic, colloidal and tartaric stabilization or ageing in oak require optimum temperatures (Bernard 1999).

4.3.1.2 N₂O Emissions

The most noxious gas in the vineyards is N_2O, considering the high greenhouse potential of this gas. It contributes to the depletion of the ozone layer in the stratosphere (Portmann et al. 2012). Research results estimate that about 50% of the vineyard pollution is generated by this gas (Wine Institute 2014). Nitrous oxide (N_2O) results naturally by nitrification/denitrification of organic fertilizers and especially synthetic nitrogen application. Ammonia by biological oxidation is transformed into nitrite (nitrification) followed by next step of nitrogen cycle, conversion of nitrite into nitrate. Nitrate ion is one of the most soluble anions in water. All these transformations are strongly influenced by temperature, humidity, pH and carbon amount in soil (Fan and Li 2010). Nitrous oxide (N_2O) release from the soil is influenced by the unstable carbon stock due to the incorporation of green fertilizer, weeds or stubble into the soil and more nitrates combined with soil moisture. A higher amount of labile C and NO_3 from soil leads to a higher amount of N_2O emission in the atmosphere (Butterbach-Bahl and Dannenmann 2011).

On the other hand, the nitrogen application as fertilizer increases the amount of sequestered carbon in the canopy and vine wood, which through leaves and pruning wood is added into soil year after year (Bouwman et al. 2002). The largest amount of nitrogen (nearly 75%) is stored in vine roots, trunks and canes (Bates et al. 2002).

In order to decrease nitrogen emissions, it is necessary to apply fertilizers in the optimum quantity during the active growth of the roots (before bud break and after harvest, respectively) correlated with the optimum temperature and humidity in the soil. Opinions concerning the contribution of soils to N_2O emissions are divided. On the one hand, soil tillage increased emissions as a result of intensive denitrification process (higher NO_3 accumulation), and on the other, cover crops increase emissions after extraction of larger amounts of nitrogen from the soil that is no longer decomposed by microbial biomass (Suddick et al. 2011). Relative equilibrium was found by Garland et al. (2011) in Mediterranean vineyards between soil emissions and no-tillage system.

CO_2 Emissions

Soil carbon is related to soil quality and is strongly linked to the nitrogen cycle, both being components of organic matter in the soil. It has been found that carbon stocks increase with depth due to the addition of organic matter in the deeper soil layers during soil tillage (Suddick et al. 2010). Soil organic carbon ensures fertility and soil health (Smith et al. 2008). As more CO_2 is stored in the soil as biomass or organic matter, lower the concentration of this gas in the atmosphere (Alvaro-Fuentes et al. 2008). The carbon stock in the soil depends on the climate, soil texture, land use and vegetation. Soil covered with natural vegetation accumulates a higher amount of carbon compared to those in which frequent and deep tillage are performed (Krull et al. 2001); CO_2 is released from the soil when organic matter is decomposed and taken up by plants. Direct emissions of GHG are produced mainly from field tractors and equipment, in wineries by diggers, forklifts, water heaters, bottling halls, etc., and electric power consumption. Grapevine is one of the perennial crops that are preserved for decades and can act as a carbon sink through the wood that grows continuously and through the pruning debris which can remain on the ground. Debris and leaves from soil help to increase the carbon stock for a long period of time (Johnston et al. 2017).

Grapevine maintenance and cover crops can increase the amount of organic matter in the soil. Biomass from cover crops or other sources decomposes over time with the release of CO_2. Carbon sequestration in soil is a long process. Climate and soil play a major role in storing carbon in the soil. More carbon amount in the soil is correlated with less N_2O emissions (Garland et al. 2011).

Data on GHG and carbon sequestration in vineyards are scarce, as research requires long time (3–5 years for N_2O and CH_4 and 10–20 years for C sequestration—Carlisle et al. 2010). The accuracy of the research data is relative. For example, Rugani et al. (2013) reported 22% CO_2 emission for the vegetation and packaging cycle, while Bosco et al. (2011) found 7% carbon emissions for planting stage in Italy, with many tractors passes for tillage and planting with a lot of fossil fuel consumption. More conclusively, Marras et al. (2015) specified CO_2 emission of 0.39 kg/1 kg of grapes in South Sardinia (Italy) vineyard.

The amount of carbon sequestered in the soil can be increased by using green fertilizers grown between rows, cover crops or pruning debris. This type of floor management is, however, unclear because plants are competing for nutrients and especially for water with grapevine. Very well-developed root system also leads to an increase of carbon sequestered in the soil. To limit the temperature increase in soil and decomposition of organic matter, resulting in CO_2 emissions decreases and alley-row mulch can be used. Grape berries contain great amount of carbon during fruit set and growing season. Powlson et al. (2011) specify that: "in a temperate environment, organic matter from soil, after one year sequesters only 1/3 carbon from the initial content and the remaining is released in the atmosphere".

Emissions from Wine Closures

Cork closures have a minor impact on carbon emissions (4%), being environmentally friendly and wine consumers' favourites. Some researchers even argue that these corks contribute to GHG mitigation as they are a bio-based product. Cork trees are even considered as an important reservoir for carbon storage in the soil as a result of the conversion of CO_2 into O_2 during photosynthesis and organic matter in cellulose.

According to Pereira et al. (2007) studies in cork forest near Evora from Portugal, 179 g C cm^{-2} are sequestrated annually. In Portugal, 4.8 million tonnes per year or 5% from the entire emissions of CO_2 in the country can be absorbed by the cork forests. The 1-year absorption of these greenhouse gases is equal to total emissions of 490,000 cars (Pereira et al. 2007). The accumulated thickness of all layers of cork removed from a cork tree throughout its life (about 200 years) is 3–4 times larger than a tree from which it has never been harvested. Using aluminium caps involves emission of larger amount of greenhouse gases, followed by plastic closures (10% emissions) that are made usually from recyclable plastic (Marin et al. 2007).

Winery Emissions

In Romania, there are more than 250 wine cellars. From those, 140 produce and sell bottled wine. Wineries are generating greenhouse gases during various activities. To produce the wine, energy is needed for crushing and pressing the grapes, filtering the must, for cooling or heating the fermentation tanks and finally for bottling, storing and transporting the wine (Niccolucci et al. 2008).

In wineries, part of the energy consumption (electric, fossil fuels) can be replaced to a certain extent by solar energy or other renewable sources, for increasing efficiency and mitigation of GHG emissions from lighting, fermentation tanks and filtration, of refrigerators, etc. Night-time cooling tanks, windows and large doors, for shorter gap in temperature reduction, help to reduce energy consumption by up to 15% (case study in Recas vineyards). For barrel wine maturing and sensor quality pattern, most wineries in Romania have built underground cellars that keep constant temperature throughout the year without energy consumption (Șerbulea and Antoce 2016).

Hot water used to wash bottles can be reused for washing other equipment. Recycling of wastewater from wineries can be used to irrigate vineyards with a significant reduction in energy consumption for pumping and bringing water from longer distances. Comandaru et al. (2012) studied wine life-cycle assessment

(LCA) in wine production from north-east of Romania (Iasi County) facility (75,000 hl/year wine production) to set the impact on environment of one white wine bottle. Production stages, energy and transport had significant impact on wine LCA. Winemaking has the major contribution on water consumption due to the large volume of wastewater during wine production. A lot of energy is necessary for removing from wastewater the pollutants until normal limits.

An option for GHG mitigation in wineries is to generate own electricity. For example, Carastelec winery from Salaj County (north-west of Romania) works with green energy: heating and cooling are done with heat pumps, and energy consumption is provided by solar panels; each year (without climate or pest damages), the winery produces 200,000 wine bottles from grapes harvested on 22 ha vineyard.

Stefanesti winery (Arges County, South of Romania) has 25 ha vineyards and is the only one cellar with completely energy autonomy in Romania. Each year, the winery produces 40,000 wine bottles. Electricity is provided by the 102 solar panels installed on the roof with a total of 25 kw (Fig. 4.2). The wine cellar also has a geothermal heat pump, powered by three drillings with 120 m deep. For the water required for the wine cellar, drills were made at 200 metres deep (Grigorescu 2018).

Colman and Päster (2009) calculate the carbon footprint taking into account the agrochemicals, mechanization, water for one tonne of grapes, electricity, natural gas, bottling, transport of bottled wine, etc., for the 2001 wine world production (2,668,300,000 l), and the result was 0.08% of whole GHG emissions. The amount generally not considered impressive is however equal to the emissions of about 1 million cars during 1 year (Colman and Päster 2009).

Transportation is one of the major sources of GHG in wine industry. In the recent year's flex tanks of 25,000 l, bottling and packaging near the market destination are viable alternatives, especially for table wines. Biofuel engine, electric engines for tractors, fewer passes by tractor, two tools attached to tractor (one rear,

Fig. 4.2 Stefanesti winery with 102 solar panels on the roof (*Source* Denis Grigorescu)

one front for two treatments/operations in the same time) reduce the time for tractors use and are options for the future.

Bottling

For bottling the quality wine, energy is consumed for the bottles and boxes production necessary for packaging. To decrease the carbon footprint, there is an alternative to bottling in light bottles. If the weight of a classic bottle of 0.75 cL is currently around 500 g, one lightweight glass is only 300 g (Forsyth et al. 2008).

"Lightweight" glass, as it is called by manufacturers, is made by reducing the thickness of the wall and removing that thick part that is naturally found in the bottom of the glass. Reducing the raw materials such as quartz sand and sodium carbonate led to an overall cost savings of 10%. Manufacturers of such bottles are also advised to use a larger amount of recycled glass, such as glass pellets. In 2003, a Hopland-based wine factory in California decided that their wine would be bottled in over 23 million of such bottles. Thanks to this choice, the amount of greenhouse gases in the USA decreased by more than 14%, more precisely by 2985 tonnes of CO_2 in 2003 from this factory. This greenhouse gas reduction was equivalent to planting over 70,000 trees and raising them for 10 years (Associated Press 2008).

Bag-in-box is an organic way to pack and transport wine with reduced carbon footprint (it was invented in the 1950s in the USA). This type of packaging reduces carbon footprint by 40%. Another advantage is that they are easy to handle and do not break easily. The disadvantage is short shelf-life (up to 9 months), and metallic polyester can crack. Bag-in-box or boxed wines that first appeared on the market in 1960 used for bulk wines were considered cheap at the time. This type of bag has one or more layers of cardboard, flexible and high strength (Yam 2009). Girboiu, Ostrov Domains, Budureasca, Oprisor Wine Cellars, Recas Wine Cellars, Odovidis-Jaristea or Vinarte are just some of the Romanian wineries selling PGI wines and table wines (white, rose and red), dry, semi-dry or semi-sweet, "bottled" in PET or a 3, 10 or 20 l bag-in-box.

Times have changed, and today, this type of packaging is mainly for certain varieties of young white wine. Manufacturers claim that this packaging presents the easiest way to open, compared to all currently available containers on the market. In spite of the efforts being made for global spread, this type of packaging still has the disadvantage of wine oxidation even when it is not opened. Thus, it cannot be used on an industrial scale for the long-term preservation or ageing the wines.

In conclusion, choosing the wine according to the bottling type remains to the consumer. However, it should be known that if wine will be consumed in a year or two, alternative packaging should be used while wine for ageing and storage should be bottled in glass containers (Penela et al. 2009).

In recent years, for cheaper table wine, bottling in plastic (PET) bottles from recyclable material (from 0.25 to 1.5 l) has been tried, to replace the glass bottles and make the transport easier and safer. PET helps to protect against colour oxidation over long periods of time and minimize temperature differences. They are

100% recyclable and 90% lighter than a traditional bottle, and they reduce transport costs and the amount of fuel used by trucks to deliver. However, consumers do not appreciate plastic packaging primarily for environmental reasons, but also the quality of wine that oxidizes much faster than in glass bottles (Imkamp 2000).

One of the most viable alternatives for wine bottling is Tetrapackaging. Tetra pack containers are from paper and weigh 40 grams compared to glass bottles weighing up to 700 grams. The production of these packages is done with about 92% less raw materials and 54% less energy. This means 80% less greenhouse gas emissions and 60% less solid waste. Additionally, these containers can be easily stacked and are resistant to transport and storage because they do not break (Borg 2013).

Unfortunately wines bottled in Tetra Pak, "lightweight" bottles, etc., are for retail selling because the consumer has the impression that the wine is of poor quality and does not realize that the price is lower due to the low cost of the packaging. The bag-in-box has 80% less carbon footprint than glass packaging (the entire production chain from vineyard to the wine bottle). Because producers want to sell their wine, packaging decision is determined by the market and consumer preferences rather than by the winemakers (Colman and Päster 2007).

Globally, 2.7 billion people are affected by water scarcity for at least 1 month each year (Degefu et al. 2018). Water footprint in food is high (e.g. 1 kg of beef requires around 15 thousand litres of water, Gerbens-Leenes et al. 2013). Winemaking industry is not an exception. Ene et al. (2013) evaluated the water footprint of a 750 cl bottle of wine produced in a medium-sized wine cellar in Romania, based on the production chain diagram representing current emissions and environmental impacts. The results of this study indicated that nearly 99% of the total water footprint is related to the use of water in the supply chain. The three water footprints are: green—water from precipitation stored in the root zone and is incorporated by plants, evaporated or transpired; blue—comes from groundwater or surface resources and is incorporated in products or evaporated; grey—is fresh water necessary to assimilate pollutants for meeting water quality standards (Bonamente et al. 2016). GHG emissions decreasing by water, raw materials or energy savings in the winemaking industry from Romania are an actual goal. Several practices are already applied in recent years (vineyards floor management, pest control by novel technologies, wineries waste treatment and monitoring, or increase water resources use efficiency).

4.3.1.3 Ageing Wine in Barrels

Wood barrel will never go out of winemaking. Maturing wine in barriques (225 l barrels, equivalent to 300 bottles of 0.75 l) is very popular in wine cellars that want to market the highest quality wines. The barrels of oak and acacia (for white wines) have been used since ancient times for flavour and preserving the wine quality.

About three and five barrels can be manufactured from one oak tree (depending on size). An oak tree of 100 years old and 20 m height has leaves that cover 1,600 m^2 area, and produces 12.8 kg/day or 4.672 kg/year of O_2 that is necessary for 11 peoples and absorb about 2,265 kg of CO_2 (del Alamo-Sanza and Nevares 2015).

The wine ages in oak barrels for approximately 3 months to 2 years, depending on the wine and the number of barrel use. A barrel can be used 3–4 times. Subsequently, the wood tannins will be exhausted and the pores of the wood get clogged. Between the uses, the barrel is hygienized by repeated rinsing with hot water and sulfation. An interesting observation has been made that Romanian wines improve their aroma and taste qualities in American or French oak barrels, while South African wines are improved in Romanian oak barriques.

Wine Distribution

Wine is produced usually in specific viticultural regions and must be transported to the warehouses, markets and wine drinkers. The transport chain depends on distance (trucks, rails and ship or air cargo) and can thus have a large carbon footprint. Air cargo has the bigger impact concerning carbon emissions (11 times more than 460 shipping, followed by trucking 5 times), according to the research results of Colman and Päster (2007).

4.4 Vineyard Floor Management

Grapevine grows on the same land for at least 30–35 years, and it is an intensive labour crop and over the year requires significant soil tillage. Therefore, vineyard soils are generally anthropic soils, poorly structured, low in humus and capillary porosity, with severe erosion on sloping lands, reduction of soil organic matter; soil compaction after repeated tractor and equipment passing in wet periods; imbalance of mineral nutrition, which generates a high sensitivity to pathogens (diseases, pests, frost, drought, etc.). Correct tillage in the best moment is very important for humus preservation, nutrients accessibility, weed control, chemical and biological activity (Duda et al. 2014). Nearly 60% of the vineyard area is covered by middle rows, an unproductive field but with major impact on grapevine yield and quality. Vineyard floor can be managed by tillage (bare soil), herbicides, cover crops, green manure, mulch, etc. (Dobrei et al. 2014). To avoid soil structure degradation, tractor and additional equipment should be used such that several different management activities are carried out simultaneously. Alleyways with cover crops as buffer between tractors tires and soil, mulching, green manure are only few possibilities to reach higher yield and production and to have a healthy soil.

4.4.1 Bare Soil Tillage

Cover crops are used in vineyards since ancient times, but few data are available about their influence on soil carbon cycle. Until the early 1980s, bare soil was the traditional floor management in viticulture (Pool et al. 1991). Bare soil advantages are more efficiency of water in the soil, increasing the amount of nitrogen, mobile phosphorus and potassium in the soil and enhancing the photosynthesis process in the vine leaves (Murisier 1981). Disadvantages include the acceleration of the humus degradation; destruction of soil structural aggregates as a result of four to five tractor passes per year; increasing soil erosion process; large volume of dust that favours mite development and air pollution; high fuel consumption and expenditure for tillage works (Martinson 2006).

Different floor managements in Burgundy vineyards from the West of Romania, including bare soil under-vines and middle rows or cover crops between rows, confirm other research results concerning cover crops influence on chemical and physical soil traits, vine vigour or crop load and weed control improvement (Dobrei et al. 2015). Nitrogen infiltrates more slowly in the soil with real effect on canopy and grape yield. Soil organic matter is a major component of soil, with extremely important contribution of water and nutrients for plant nutrition and for soil carbon sequestration. Minimum tillage on several soil types (phaeozem, haplic luvisols, chernozem and molic fluvisol) from north-west of Romania increased the organic matter from 0.8 to 22.1% and water stability of soil aggregates (WSA) up to 13.6 from 1.3% in the first 30 cm of soil, compared with bare soil system (Moraru and Rusu 2010).

Extending these results to 50% of the Romanian arable land, it was estimated that 6.9 million tonnes/year of carbon could be stored if minimum tillage is performed. Furthermore, no-tillage or moderate tillage on argic–stagnic faeoziom soil (north-west of Romania) influences the time and the amount of carbon sequestration. Results after 3 years of observations show the effect of soil tillage on daily average soil respiration as follows: no-tillage had the lowest influence (315–1914 mmoli m^{-2} s^{-1}) and 318–2395 mmol m^{-2} s^{-1} for moderate tillage system, respectively, compared with conventional tillage (321–2480 mmol m^{-2} s^{-1}) (Moraru and Rusu 2013). Tractors and equipment used for minimum tillage on argic–stagnic faeoziom in Jucu Experimental Station (north-west of Romania) increased the GHG emission twice compared with no-tillage soils. The smallest rate of CO_2 emission in minimum soil tillage was 1929 ppm, and the highest rate was 8901 ppm. When no-tillage was adopted, the CO_2 concentrations registered were between 1443 ppm and 4880 ppm/day (Marian et al. 2013).

Relation between soil, climate, grapevine and variety influence on wine character is different in each viticultural region. Management of vineyard floor is particularly important in the wine production chain. Soil moisture has a strong impact on grape yield and quality during growing season. In Valea Călugărească vineyards, Fetească regală grapevine located on mollic reddish-brown soil, floor management was evaluated during 2012–2013. Soil moisture was measured, and comparisons among bare soil, straw mulch across vine rows and alleyways (10 cm layer), mulching with pomace compost alleyways (10 cm layer) and minimum tillage were made. During both years of experiment, soil moisture in the first 60 cm was normal and quite equal in the early growing season (April–May), but in summer and autumn time, mulching system plots had positive influence on water evaporation and soil moisture (Serdinescu et al. 2014).

Minimum tillage increases significantly the humus amount in soil by 0.8–22.1% especially on vertic preluvosoil. Hydro-stability of macroaggregates and organic carbon is positively influenced by minimum tillage from 1.3 to 13.6% in the first 30 cm of soil compared with the conventional tillage. Both humus amount and soil structure contribute to increasing soil fertility and have positive influence on soil permeability and water storage in soil as groundwater storage (Duda et al. 2014). Similar results have been observed on Somes Plateau argic faeoziom soils (north-west of Romania), when no-tillage, minimum tillage and conventional tillage were compared for soil respiration. In no-tillage system, the lowest soil respiration ($315–1914$ mmol m^{-2} s^{-1}) was found. CO_2 release from soil in minimum tillage was ($318–2395$ mmol m^{-2} s^{-1}), while in conventional tillage, production of CO_2 was the highest ($321–2480$ mmol m^{-2} s^{-1}). The CO_2 production was maximum in autumn ($2141–2350$ mmol m^{-2} s^{-1}) and in late spring ($1383–2480$ mmol m^{-2} s^{-1}). After 3 years, humus amount in soil increased by 0.64% when no-tillage was applied, followed by minimum tillage system with 0.41% humus level in soil (Rusu et al. 2016).

4.4.1.1 Grass Strips Alleyways

Soil is subjected to various degradation processes. Some of these are specific to viticulture: water and wind erosion or soil tillage; compaction; decreasing the amount of organic carbon in soil and soil biodiversity; soil salinisation and sodification; soil contamination with heavy metals and pesticides or excessive amounts of nitrates and phosphates. On clay soils, repeated tractors passing decreases soil porosity in the vine root area from middle rows. The fine soil particles by leaching agglomerate the deep layer resulting in a long-lasting compacting with adverse consequences on the vine vigour and productivity. Symptoms are yellowing or redness of the leaves associated with abnormal deformation of the leaves of vines (Valenti et al. 2002).

Grass strip alleyways and bare soil under-vines in vineyards are a common soil management system and are efficient in areas with more than 600 mm annual rainfall. Weed control on vine row is done by manual hoeing or by repeated application of residual or foliar herbicides; this system is recommended for increasing land slope stability (Murisier and Sbeuret 1986). The advantages of this system application are: reduction of erosion on sloping lands; avoiding hardpan formation; increasing the input of organic matter in the soil; improving soil structure and porosity; avoiding nitrogen runoff and leaching throughout the year; better water infiltration (Colungati and Cattarossi 2013).

Constant tillage exposes the soil to erosion as many of the vineyards are located on slopes with mild or medium slopes. On these lands, the main problem is not the landslides, but the gradual and constant transfer soil which decreases its fertility, causing a "slip" of the roots that are thus forced to expand around to find the necessary nutrients and water in the soil (Kaspar et al. 2001). Grass strips are an alternative for vineyards' middle row soil protection. In organic viticulture, this is the first choice in vineyards floor management for increasing soil fertility and for restoration of degraded or weed-infested soil. Water and wind erode the soil by about 2.5 cm per year (Goulet et al. 2004). Grass strips protect the soil by avoiding leaching, compaction and erosion due to tractors and other equipment traffic (middle rows grass strips favour iron and phosphorus absorption, contribute to less mobile minerals alteration, reduce the rot attacks, make easier the access to the vines, and reduce the maintenance costs (Eynard and Dalmasso 2004).

Grass strips in the middle rows contribute to the improvement of grape quality, by increasing the amount of sugars in berries, vine vigour and grape yield. This system of floor management in vineyards mainly protects the soil from erosion (Goulet et al. 2004). For example, in the Rheingau (Germany) wine-growing region, on slopes ranging from 10 to 32%, grass strips reduced water running after heavy rains up to 1.8% from rainfall volume compared to 50% on the bare soil (Krull et al. 2001). The amount of soil lost by erosion in vineyards with grass strips is insignificant (3.1 kg/ ha), while on bare soil it is between 30 and 100 t/ha, depending on the degree of slope (Emde 1990). Grass strips enhance soil aggregates stability in the first 10–15 cm (Condei and Ciolacu 1991). In vineyards from many countries (Austria, Germany, Switzerland, Italy, France, Romania, etc.), middle rows are covered with mixtures of plant species. This floor management system is recommended in regions with annual rainfall of 600–700 mm, of which at least 250 mm occur between May and August (Valenti et al. 2002).

Well-structured soils store a lot of water, air, heat and nutrients, ensuring favourable conditions for vine growing and fruit set. Vineyards' middle raw soil maintenance with permanent grass cover crops improves the structure and physical properties of the soil due to the organic matter addition and increasing microbial biomass through the biological activity (Bandici 2011; Dobrei et al. 2016).

Under high or too low soil moisture, tractors' passing has negative impact on soil physical and chemical properties (Dobrei et al. 2008). Besides financial advantages, the permanent grass strips also provide advantages such as protection against soil erosion and degradation, allowing phytosanitary treatments in favourable stages, reduced water loss. (Dobrei et al. 2009).

The extended grass roots contribute to the soil loosening by adding organic matter transformed into humus that contributes to the activation of microbial life. Both the root mass and the above-ground plant contribute to humus restoration, the beneficial effect of dead plants being influenced by their chemical composition and diffusion in the soil (Bernaz and Dejeu 1999).

Grasses' extent root system prevents soil erosion on sloping lands and is exposed to wind damage. Perennial plants can absorb nitrogen from the soil, but unlike legumes can only contribute to soil improvement by supplying biomass. Annual grasses such as rye, oats, barley or triticale sown in autumn are mowed or buried in spring to protect against frost. Therefore, the soil absorbs more heat over the day and releases it at night. Stubble left after mowing competes with weeds and contributes to decrease dust level and soil compaction after multiple tractors passages (Christensen 1971). However, besides advantages, natural vegetation or seeded plants can become competing for water and nutrients with vines. This competitiveness depends on the climatic conditions (especially the rainfall amount), grapevine requirements for water, the water absorption capacity of the cover crops species and the type of soil (Sicher et al. 1993). The disadvantage of natural vegetation is also derived from slowly and unevenly growing and clear space for weeds (Sicher and Dorigoni 1994).

Alternate Clean Cultivated and Grass (Legumes) Strips Alleyways

Floors managed with alternate clean cultivated and grass (and legumes) strips alleyways are recommended in vineyards with at least 350 mm rainfall during summer season (April–October) (Piţuc 1989). This system limits water runoff, soil erosion, increases the input of organic matter into the soil, allows tractors passing on wet weather, and reduces fuel consumption and the manual work (Condei and Ciolacu 1991). Usually after spring tillage, short species such as lawn grass (Lolium perenne—12–14 kg/ha), white clover (Trifolium repens) or species with spread roots like oilseed radish (R. sativus var. oleiferus or Raphanus sativus) are seeded. These plants can improve the soil structure, water drainage and fast root development (Bernaz and Dejeu 1999).

Grass mowed during the growing season is left on the ground as mulch. After 8–9 years, the grass strips are dissolved by tillage and the alleyways are changed with bare soil. Growing grass between vine rows for long period is not advisable because the soil becomes compacted, as emphasized by multiple tractor passing which increases soil stress. Therefore, the soil is loosened and reseeding is recommended every 4 or 5 years (Bernaz and Dejeu 1999). Once the symbiosis process of the legume plants starts, elements like nitrogen, phosphorus, potassium, iron and other minerals are released and absorbed by the soil. The vigorous root system that can reach 2–3 m underground helps to loosen soil, soil oxygenation and drainage of excess water resulting from heavy rains, snow melting, etc. Under dry and high temperature period, the well-developed leaves protect the soil from the sunburn, thus avoiding dehydration.

Plant species sown as cover crops are different for each viticultural region, climate and slope. For example, in vineyards from the Galati region (south-east of Romania), on dry and medium humidity areas with slope $\leq 10\%$, alleyways were covered with green manure (mixture of common vetch—Vicia sativa (120 kg/ha) and oat (60 kg/ha)) alternate with clean cultivation alleyways. In the same vineyard have been tested alternate alleyways of green manure (grass and legumes mixture) alternate with natural vegetation (Enache 2007).

Smooth brome (Bromus inermis Leyss.) strips, 1.2-m wide, were tested in sloping vineyards from the Moldavia region for water runoff and soil erosion control. The amount of soil loss during 1 year was estimated at 1.2 m^3/ha, which is considered tolerance value. The fibrous root system contributes to reduce soil erosion and remove excess nitrogen from the soil (Enache 2007).

Peas (150–200 kg/ha), lupine bean (150–200 kg/ha), broad bean (150–200 kg/ha), soybean (150–200 kg/ha), grass legumes mixture (60 kg oat/ha + 120 kg peas/ha), rye (80–100 kg/ ha), vetch (120 kg/ha) are recommended as green manure (Enache 2007). They are fast-growing crops, with the possibility of atmospheric nitrogen fixation. Green manure is recommended in vineyards with annual rainfall over 600 mm, as well as in irrigated vineyards with large middle rows (3.0–3.6 m). Because of this, the soil is enriched in organic matter and nutrient mobility increases; excess moisture is taken by plants during the first stage of growing, and soil erosion reduces by using green mulch (Duda et al. 2014). Green manure is recommended for sandy soils. Mustard or rape on clay soils is recommended and peas on acidic soils. Lupine and clover are suggested as green manure on sandy soils. Debris is added by tillage into first 10–20 cm on sandy soils in early spring and 5–10 cm on clay soils depending on the soil type, moisture and amount of biomass. It is not advisable to incorporate green manure into the soil shortly after rain (Dobrei et al. 2016). Organic manure increases the biological activity in soil and, therefore, the fertility. About 3–50 tons/ha manure or grape pomace compost is recommended for optimum plant nutrition, yield and production after decomposition and transformation in humus by microorganisms (Vătămanu 2012).

Benefits for soil are different depending on the C/N ratio and the burring stage (young, mature, or old plants). Young plants with a low C/N ratio produce a small amount of organic matter but significantly stimulate soil biological activity (as source of minerals for microorganisms found near roots). However, mature or old plants with a high C/N ratio give an increased contribution of organic matter in soil (Moraru et al. 2015).

In vineyards from Danube terraces with rainfall over 600 mm, annually the green manure alleyways are recommended, for proper soil water balance and less soil erosion, with positive influence on organic matter in soil and less fuel consumption of 20–24 l/ha/year. Similar results have been observed on Someş Plateau argic faeoziom soils (north-west of Romania), when no-tillage, minimum tillage and conventional tillage were compared for soil respiration.

Fuel Consumption

In wine industry, fuel consumption is one of the major sources of GHG emissions. It is already known that fuel consumption represents among 25–40% from total energy input in a crop. Floor management in both alleyway and under-vines requires a lot of fossil fuel and contributes to the exposure of organic matter to microbial decomposition and consequently to the CO_2 release (Carlisle et al. 2009). Less soil tillage contributes not only to the reduction of fossil fuel consumption but also to soil erosion and to the increase in carbon, nitrogen oxide and water supply into the soil.

Most vineyards are found on hillside lands with gentle slopes (always slopes to the south are preferred), sheltered from winds and warm in the growing season (April–October) (Gasso et al. 2014). Fuel consumption is different depending on the region, size of vineyard, tillage system, soil type, strength and moisture, land slope or altitude (Sørensen et al. 2014). Fuel consumption was variable on different soil types from Apahida, Cluj County (Stănilă 2014). On the same soils, depending on slope deep, fuel consumption is higher up to 21% on 9 to 14° slopes compared with flat land fuel consumption (Stănilă 2014).

Soil tillage by mouldboard plough at depth between 18 and 35 cm consumes fuel from 14.61 to 20.67 l/ha (Moitzi et al. 2014). By machinery traffic control in the vineyard, and suitable tillage method, total emissions of GHG are reduced, and soil compaction and runoff decrease. Therefore, in conventional tillage fuel consumption can be decreased by less soil tillage depth or by substitute conventional with minimum or no-tillage systems. Comparing conventional tillage with minimum tillage system, fuel consumption can be decreased from 72.93 to 48.26 l/ha (Stănilă et al. 2013). Performing two or three tasks in the same pass increases efficiency and decreases fuel consumption.

4.5 Conclusions

Nowadays, the effects of climate change and especially of climatic variability are becoming obvious with disastrous consequences, mainly as a result of anthropogenic actions. The wine industry generates GHG, especially during the wine-making, bottling, preservation and transport to consumers. Fortunately, viticulturists have begun to look for solutions to reduce energy and fuel consumption, which are major causes of environmental pollution. In many vineyards, manual work is still being used for pruning and canopy maintenance over the year, but also for row under-vine tillage. In many wine-growing regions, cover crops, grass alleys or other environmentally friendly methods to increase soil fertility and provide natural fertilizers have been adopted. Night-time cooling tanks, windows and large doors, underground cellars that keep constant temperature throughout the year without energy consumption, heating and cooling done with heat pumps, energy consumption provided by solar panels, deep wells for water required in the wineries, are some of the solutions already applied in few wineries from Romania. In the recent years, bag-in-box wine and wine bottled in plastic (PET) bottle are beginning to be used on a small scale to assess consumer preferences. Manual labour is still being used in vineyards for tillage, pruning or harvest; the use of agricultural machinery is still limited, thus reducing the use of chemicals and fossil fuels. These are some of the factors that render Romania a country with the lowest GHG emissions from agriculture in the EU.

References

Alvaro-Fuentes J, Lopez MV, Arrue JL (2008) Management effects on soil carbon dioxide fluxes under semiarid Mediterranean conditions. Soil Sci Soc Am J 72:194–200

Ball DA, Parker R, Colquhoun J, Dami I (2014) Preventing herbicide drift and injury to grapes. Corvallis, OR: Extension Service, Oregon State Univ. Rep. EM8860, p 7

Bates TR, Dunst RM, Joy P (2002) Seasonal dry matter, starch, und nutrient distribution in Concord grapevine roots. HortScience 37:313–316

Bandici GHE (2011) Ecoagricultura, partea a III-a, (note de lectură), Capitolul VIII. Agroecosistemul viticol. Viticultura în contextul agriculturii biologice. http://agricultura-sustenabila.blogspot.ro/2011/04/gheorghe-emil-bandici-ecoagricultura

Bărbulescu O (2017) Potential and risks in the Romanian wine industry, Bulletin of the Transylvania University of Braşov, Series V: economic sciences, vol 10 (59), no 1, pp 193–202

Benedetto G (2013) The environmental impact of a Sardinian wine by partial life cycle assessment. Wine Econ Policy 2(1):33–41

Bernard M (1999) Une nouvelle technique: La maceration préfermentaire à la neige carbonique. Revue dés Oenologues et des Techniques Vitivinicoles et Oenologiques 92:26–30

Bernaz GH, Dejeu L (1999) Fertilizarea viilor şi întreţinerea solului în concepţie ecologică (Fertilization of vineyards and soil maintenance in ecological purpose), Ed Ceres, Bucureşti, pp 101–173

Bonamente E, Scrucca F, Rinaldi S, Merico MC, Asdrubali F, Lamastra L (2016) Environmental impact of an Italian wine bottle: carbon and water footprint assessment. Sci Total Environ 560:274–283

Borg S (2013) The green vine: a guide to west coast sustainable, organic, and biodynamic wineries, mountaineers books, p 192

Bosco S, Di Bene C, Galli M, Remorini D, Massai R, Bonari E (2011) Greenhouse gas emissions in the agricultural phase of wine production in the Maremma rural district in Tuscany, Italy. Ital J Agron 6:93–100

Bouwman AF, Boumans LJM, Batjes NH (2002) Modeling global annual N_2O and NO emissions from fertilized fields, Global Biogeochem Cycles 16. https://doi.org/10.1029/2001gb001812

Brunori E, Farina R, Biasi R (2016) Sustainable viticulture: the carbon-sink function of the vineyard agro-ecosystem. Agric Ecosyst Environ 223:10–21. https://doi.org/10.1016/j.agee.2016.02.012

Butterbach-Bahl K, Dannenmann M (2011) Denitrification and associated soil N_2O emissions due to agricultural activities in a changing climate. Curr Opin Environ Sustain 3:389–395. https://doi.org/10.1016/j.cosust.2011.08.004

Carlisle E, Smart DR, Browde J, Arnold A (2009) Carbon footprints of vineyard operations. Pract Winery Vineyard 2009:15–21

Carlisle E, Smart D, Williams LE, Summers M (2010) California vineyard greenhouse gas emissions: assessment of the available literature and determination of research needs, pp 8–34. www.sustainablewinegrowing.org

Carlier LI, Rotar I, Vidican R (2009) Storage of carbon dioxide: favorable for the crops production and the environment (Fixarea dioxidului de carbon: favorabilă pentru producţia culturilor agricole şi pentru mediu). BioFlux, ProEnvironment, pp 73–83

Carlton JS, Perry-Hill R, Huber M, Prokopy LS (2015) The climate change consensus extends beyond climate scientists. Environ Res Lett 10:094025

Christensen P (1971) Modified sod cover cropping in vineyards. Blue Anchor 48(3):22

Colman T, Päster P (2007) Red, white and "green": the cost of carbon in the global wine trade. Amer Assoc Wine Econ 9:1–20. https://doi.org/10.1080/09571260902978493

Colman T, Päster P (2009) Red, white, and "green": the cost of greenhouse gas emissions in the global wine trade. J Wine Res 20(1):15–26. https://doi.org/10.1080/09571260902978493

Comandaru IM, Bârjoveanu G, Peiu N, Ene SA, Teodosiu C (2012) Life cycle assessment of wine: focus on water use impact assessment. Environ Eng Manage J 11(3):533–543

Colungati G, Cattarossi G (2013) L'inerbimento del vigneto come tecnica conservativa del terreno; Gestione del suolo vitato e tutela del territorio; Le possibilità di applicazione negli ecosistemi settentrionali e in quelli meridionali, VQ, vite, vino e qualita. http://www.vitevinoqualita.it/gestione-del-suolo-vitato-e-tutela-del-territorio/

Condei G, Ciolacu M (1991) L'approche ecologique du systeme entegre d'entretien du sol en plantations viticoles intensives. III Symp. intern. sur la non-culture de la vigne et les autres techniques d'entretien des dols viticoles, Montpellier, pp 289–296

del Alamo-Sanza M, Nevares I (2015) Wine maturation: a dynamic evaluation of the oxygen transfer rate in oak barrels. Wine Viticult J 30:26

Degefu DM, Weijun H, Zaiyi L, Liang Y, Zhengwei H, Min A (2018) Mapping monthly water scarcity in global transboundary basins at country-basin mesh based spatial resolution. Sci Reports 8:1–10. https://doi.org/10.1038/s41598-018-20032-w

Dobrei A, Sala F, Kocis E, Savescu I (2008) Soil maintenance systems influence upon yield and quality in case of some vine varieties in BuziasSilagiu viticultural center, scientific papers, vol 40, no 2. Agroprint, Timişoara, pp 59–62

Dobrei A, Ghiţă A, Cristea T, Sfetcu A (2009) The influence of soil maintenance systems on vigor, quantity and production quality at some grape varieties for wine. J Horticult Forest Biotechnol 13:197–200. Agroprint, Timişoara

Dobrei A, Dobrei AG, Sala F, Nistor E, Mălăescu M, Dragunescu A, Cristea T (2014) Research concerning the influence of soil maintenance on financial performance of vineyards. J Hortic, For and Biotech 18(1):156–164. www.journal-hfb.usab-tm.ro

Dobrei A, Nistor E, Sala F, Dobrei A (2015) Tillage practices in the context of climate change and a sustainable viticulture. Not Sci Biol 7(4):500–504. https://doi.org/10.15835/nsb.7.4.9724. Print ISSN 2067-3205; Electronic 2067-3264

Dobrei A, Dobrei A, Nistor E, Stanciu S, Moatar M, Sala F (2016) Sustainability of grapevine production through more efficient systems of soil maintenance and agro-biological indicators. In: Raupelienė A (ed) Proceedings of the 7th international scientific conference rural development 2015. ISSN 1822-3230/eISSN 2345-0916e, ISBN 978-609-449-092

Duda BM, Rusu T, Bogdan I, Pop AI, Moraru PI, Giurgiu RM, Coste CL (2014) Considerations regarding the opportunity of conservative agriculture in the context of global warming. Res J Agric Sci 46(1):210–217

Emde K (1990) Oberflaechenabfluss und Bodenerosionmessungen auf Weinbergstandrten im Rheingau (Deutschland). Act of VIII: Int. Kolloquium über Begruenung im Weinbau, Keszthely (Hungary), pp 163–169

Enache V (2007) Aspects regarding antropic factor involvement about erosional development fighting process on wine-growing fields in context of a durable agriculture. In: de la Brad II (ed) Scientific papers horticulture, vol 50. Iasi, Romania, pp 627–632

Ene SA, Teodosiu C, Robu B, Volf I (2013) Water footprint assessment in the winemaking industry: a case study for a Romanian medium size production plant. J Clean Prod 43:122–135

Eynard I, Dalmasso G (2004) Viticoltura moderna: manuale pratico: evolutione della viticoltura, noziioni, Hoepli (ed), pp 473–475. ISBN 88-203-1768-0

Fan XH, Li YC (2010) Nitrogen release from slow-release fertilizers as affected by soil type and temperature. Soil Sci Soc Amer J 74:1635–1641. https://doi.org/10.2136/sssaj2008.0363

FAO (2017) Soil organic carbon: the hidden potential. Food and Agriculture Organization of the United Nations Rome, Italy

Forsyth K, Oemcke D, Michael P (2008) Greenhouse gas accounting protocol for the international wine industry (The Wine Institute of California, New Zealand Winegrowers, Integrated Production of Wine South Africa and the Winemakers Federation of Australia Report)

Fraga H, Malheiro AC, Moutinho-Pereira J, Santos JA (2012) An overview of climate change impacts on European viticulture. Food Energy Secur 1:94–110

Garland GM, Suddick E, Burger M, Horwath WR, Six J (2011) Direct N_2O emissions following transition from conventional till to no-till in a cover cropped Mediterranean vineyard (Vitis vinifera). Agric Ecosyst Environ 144(1–2):423–428

Garwood EA, Tyson KC, Clement CR (1977) A comparison of yield and soil conditions during 20 years of grazed grass and arable cropping. Grassland Res Inst Tech Rep 21:89

Gasso V, Oudshoorn FW, Sørensen CAG, Pedersen HH (2014) An environmental life cycle assessment of controlled traffic farming. J Clean Prod 73:175–182

Gerbens-Leenes PW, Mekonnen MM, Hoekstra AY (2013) The water footprint of poultry, pork and beef: a comparative study in different countries and production systems. Water Res Industr 1–2:25–36. https://doi.org/10.1016/j.wri.2013.03.001

Goode J, Harrop S (2011) Authentic wine: toward natural and sustainable winemaking, Univ. of California Press Ltd., Chap 12: The carbon footprint of wine. ISBN 978-0-520-26563-9, pp 219–234

Goulet E, Dousset S, Chaussod R, Bartoli F, Doledec AF, Andreux F (2004) Water-stable aggregates and organic matter pools in a calcareous vineyard soil under four soil-surface management systems. Soil Use Manag 20:318–324

Grigorescu D (2018) The wine made by Rebreanu from love to his wife Fanny, "reanimated" to Ştefăneşti. The story of the only autonomous energetic cellar in Romania (Vinul făcut de Rebreanu din dragoste pentru soţia Fanny, "resuscitat" la Ştefăneşti. Povestea singurei crame din România complet autonome energetic), Adevarul: http://adevarul.ro/locale/pitesti/vinul-facut-rebreanu-moartea-sotiei-fanny-refabricat-decenii-uitare-trei-tineri-produc-singura-crama-romania-complet-autonoma-energetic-1_5a8957e7df52022f75af93d7/index.html

Holdren JP (2018) Meeting the climate-change challenge: what do we know? what should we do? Watson Distinguished Speaker Series, Institute at Brown for Environment and Society Watson Institute for International and Public Affairs, Brown University, February 15. http://whrc.org/wp-content/uploads/2018/02/2018-02-15_climate-change_challenge_Brown_JPH.pdf

Iannone R, Miranda S, Riemma S, De Marco I (2016) Improving environmental performances in wine production by a life cycle assessment analysis. J Clean Prod 111:172–180

Imkamp H (2000) The interest of consumers in ecological product information is growing—evidence from two German surveys. J Consum Policy 23:193–202

Irimia LM, Patriche CV, Roşca B (2017) Climate change impact on climate suitability for wine production in Romania. Theoret Appl Climatol 1–14. https://doi.org/10.1007/s00704-017-2156-z

Johnston AE, Poulton PR, Coleman K, Macdonald AJ, White RP (2017) Changes in soil organic matter over 70 years in continuous arable and leyarable rotations on a sandy loam soil in England, 68(3):305–316

Kaspar TC, Radke JK, Laflen JM (2001) Small grain cover crops and wheel traffic effects on infiltration, runoff, and erosion. J Soil Water Conserv 56:160–164

Krull E, Baldock J, Skjemstad J (2001) Soil texture effects on decomposition and soil carbon storage. In: Nee workshop proceedings, pp 103–110

Marian R, Rusu T, Braşovean I, Milăşan A, Ugruţan F (2013) Effect of tillage practices on the moisture, temperature and soil carbon dioxide flux. ProEnvironment 6:227–232

Marin A, Jorgensen E, Kennedy J, Ferrier J (2007) Effects of bottle closure type on consumer perceptions of wine quality. Am J Enol Viticult 58:182–191

Marras S, Masia S, Duce P, Spano D, Sirca C (2015) Carbon footprint assessment on a mature vineyard. Agric Forest Meteorol 214–215:350–356. https://doi.org/10.1016/j.agrformet.2015.08.270

Martinson TE (2006) Sustainable viticulture in the North-East, A publication of the Finger Lakes, Lake Erie, and Long Island Regional Grape Programs, Newsletter, p 4

Moitzi G, Wagentristl H, Refenner K, Weingartmann H, Piringer G, Boxberger J, Gronauer A (2014) Effects of working depth and wheel slip on fuel consumption of selected tillage implements. Agric Eng Int CIGR J 16(1):182–190. http://cigrjournal.org/index.php/Ejounral/article/view/2661

Moraru PI, Rusu T (2010) Soil tillage conservation and its effect on soil organic matter, water management and carbon sequestration. J Food Agric Environ 8(3–4):309–312

Moraru PI, Rusu T (2013) Effect of different tillage systems on soil properties and production on wheat, maize and soybean crop. Int J Biol Biomol Agric Food Biotechnol Eng 7(11):1027–1030

Moraru PI, Rusu T, Guş P, Bogdan I, Pop AI (2015) The role of minimum tillage in protecting environmental resources of the Transylvanian plain, Romania. Roman Agric Res 32:127–135

Murisier F (1981) La culture de la vigne en banquettes: La situation en Suisse romande. Revue Suisse Viticolture, Arboricolture, Horticolture 13(2):77–82

Murisier F, Sbeuret E (1986) L'enherbement des sols viticoles. Revue suisse Vitic Arboric Hortic 18:291–294

Niccolucci V, Galli A, Kitzes J, Pulselli RM, Borsa S, Marchettini N (2008) Ecological footprint analysis applied to the production of two Italian wines. Agric Ecosyst Environ 128:162–166

Penela AC, García-Negro MdC, Quesada JLD (2009) A methodological proposal for corporate carbon footprint and its application to a wine-producing company in Galicia. Spain Sustain 1 (2):302–318

Pereira JS, Correia AP, Mateus JA, Aires LMI, Pita G, Pio C, Andrade V, Banza J, David TS, Rodrigues A, David JS (2007) O sequestro de carbono por diferentes ecossistemas do Sul de Portugal" Seminário no Auditório da Culturgest, em Lisboa a 28 de Março, policopiado, p 11

Piţuc P (1989) Înierbarea alternativă a intervalelor dintre rânduri în plantaţiile viticole din nord-estul Moldovei. Revista de Hoticultură, nr 9

Pool R, Dunst R, Senesac A, Kamas J (1991) I. Choosing a weed management program, Grape Facts, vol 1 no 1. New York State Agricultural Experiment Station. http://hdl.handle.net/1813/17494

Portmann RW, Daniel JS, Ravishankara AR (2012) Stratospheric ozone depletion due to nitrous oxide: influences of other gases. Phil Trans R Soc B 367:1256–1264. https://doi.org/10.1098/rstb.2011.0377

Powlson DS, Whitmore AP, Goulding KWT (2011) Soil carbon sequestration to mitigate climate change: a critical re-examination to identify the true and the false. Eur J Soil Sci 62:42–55. https://doi.org/10.1111/j.1365-2389.20, 10.01342.x

Robertson GP (1993) Fluxes of nitrous oxide and other nitrogen trace gases from intensively managed landscapes: a global perspective. In: Harper LA, Mosier AR, Duxbury JM, Rolston DE (eds) Agricultural ecosystem effects on trace gases and global climate change, pp 5–108. American Society of Agronomy, Madison, WI, USA

Rugani B, Vázquez-Rowe I, Benedetto G, Benetto E (2013) A comprehensive review of carbon footprint analysis as an extended environmental indicator in the wine sector. J Clean Prod 54:61–77. https://doi.org/10.1016/j.jclepro.2013.04.036

Rusu T, Moraru PI, Bogdan I, Pop IA (2016) Effects of tillage practices on soil carbon and soil respiration. Geophys Res Abstracts 18, EGU2016-3300

Scrucca F, Bonamente E, Rinaldi S (2018) Chapter 7: carbon footprint in the wine industry, in environmental carbon footprints: industrial case studies, Subramanian Senthilkannan Muthu (eds), pp 161–190. Elsevier, ISBN: 978-0-12-812849-7

Şerbulea M, Antoce AO (2016) Study of the influence of aging in different barrels on Shiraz wines, Scientific Papers. Series B, Horticulture. vol LX, pp 109–116. Print ISSN 2285-5653, CD-ROM ISSN 2285-5661, Online ISSN 2286-1580, ISSN-L 2285-5653

Şerdinescu A, Pîrcălabu L, Fotescu L (2014) Influence of soil maintenance systems and fruit load on grapes quality under drought conditions. In: Scientific papers—series b, horticulture, no 58, pp 201–204, ref 6

Sicher L, Dorigoni A, Altissimo A (1993) La gestione del suolo in fruttiviticoltura attraverso la pratica dell'inerbimento. Bollettino ISMA, nuova serie, anno I, n 2:36–50

Sicher L, Dorigoni A (1994) Come gestire l'inerbimento. Terra e Vita 14:79–82

Smith R, Bettiga L, Cahn M, Baumgartner K, Jackson LE, Bensen T (2008) Vineyard floor management affects soil, plant nutrition, and grape yield and quality. Calif Agric 62(4):184–190

Sørensen CG, Halberg N, Oudshoorn FW, Petersen BM, Dalgaard R (2014) Energy inputs and GHG emissions of tillage systems. Biosys Eng 120:2–14

Stănilă S, Drocaş I, Molnar A, Ranta O (2013) Studies regarding comparative fuel consumption at classical and conservation tillage. ProEnvironment Promediu, 199–202

Stănilă S (2014) Aspects regarding on fuel consumption at tillage on slopes. Agric Agric Pract Sci J 1–2(89–90):53–58

Suddick E, Scow KM, Horwath WR, Jackson LE, Smart DR, Mitchell JP, Six J (2010) The potential for California soils to sequester carbon and reduce greenhouse gas emissions: a holistic approach. In: Advances in agronomy, vol 107. Academic Press, Burlington, pp 123–162

Suddick EC, Steenwerth K, Garland GM, Smart DR, Six J (2011) Discerning agricultural management effects on nitrous oxide emissions from conventional and alternative cropping systems: a California case study. In: Guo L, Gunasekara AS, McConnell LL (eds) Understanding greenhouse gas emissions from agricultural management. American Chemical Society, Washington, D.C., USA, pp 203–226

Toti M, Ignat P, Voicu V, Mocanu V, Stănilă AL (2015) Features of soil cover in different terroir units from Romania. Res J Agric Sci 47(3)

Tamas A (2017) The design of the Romanian wine imports and exports using the gravity model approach. Int J Bus Manage V(2). https://doi.org/10.20472/bm.2017.5.2.005

The Associated Press (2008) California winery sees clear benefits in lightweight glass. Nov 2. https://www.chicoer.com/2008/11/02/california-winery-sees-clear-benefits-in-lightweight-glass/

Toscano G, Riva G, Duca D, Pedretti EF, Corinaldesi F, Rossini G (2013) Analysis of the characteristics of the residues of the wine production chain finalized to their industrial and energy recovery. Biomass Bioenerg 55:260–267

Valenti L, Divittini A, Gallina PM, Zerbi P (2002) Inerbimento del vigneto con specie auto riseminanti. L'Informatore Agrario 5:69–72

Yam KL (2009) Encyclopedia of packaging technology, Wiley. ISBN 978-0-470-08704-6

Zomer RJ, Bossio DA, Sommer R, Verchot LV (2017) Global sequestration potential of increased organic carbon in cropland soils, scientific reports, Art 15554, vol 7, pp 1–8. https://doi.org/10.1038/s41598-017-15794-8

Chapter 5
Agricultural Cropping Systems in South Africa and Their Greenhouse Gas Emissions: A Review

Mphethe Tongwane, Sewela Malaka and Mokhele Moeletsi

Abstract South Africa is a major emitter of greenhouse gases (GHGs) and accounts for 65% and 7% of Africa's total emissions and agricultural emissions, respectively. South Africa has a dual agricultural economy, comprising a well-developed commercial sector and subsistence-oriented farming in the rural areas. The country has an intensive management system of agricultural lands. Agriculture, forestry and other land use sector is the second largest producer of GHG emissions in the country with 12% of the national total emissions. This review presents characteristics of GHG emissions from crop management in South Africa. It establishes trends of emissions from data collated from the literature. Main sources of GHG emissions from cropping systems in South Africa are maize, sorghum, wheat and sugarcane productions. Although the emissions from the application of synthetic nitrogen (N) fertiliser to agricultural land show a slight decrease with time, this remains the main sources of emissions from cropping systems in the country. On the other hand, national emissions from urea fertiliser are increasing. Emissions from management of crop residues are low. Conversion of land to croplands is a net source of CO_2 emissions in South Africa. Lack of investment in biofuels and production preference given to previously disadvantaged farmers has slowed their uptake. All stakeholders have to contribute actively to address the current poor status of linkages between agricultural research and policy in the country in order to reduce the current growth of agricultural emissions.

Keywords Crop management · Agricultural lands · Field crops
Biofuels · Mitigation strategies

M. Tongwane (✉) · S. Malaka · M. Moeletsi
Agricultural Research Council – Institute for Soil, Climate and Water, Private Bag X79,
Pretoria 0001, South Africa
e-mail: tongwanem@arc.agric.za

M. Moeletsi
Risk and Vulnerability Assessment Centre, University of Limpopo, Private Bag X1106,
Sovenga 0727, South Africa

© Springer Nature Singapore Pte Ltd. 2019
N. Shurpali et al. (eds.), *Greenhouse Gas Emissions*, Energy, Environment,
and Sustainability, https://doi.org/10.1007/978-981-13-3272-2_5

5.1 Introduction

Anthropogenic greenhouse gas (GHG) emissions are mainly driven by increasing demand for human food, economic activity, lifestyle, energy use, land use patterns, technology and climate policy (Thornton 2010; Vermeulen et al. 2012; FAO 2014; IPCC 2014). The population in Africa has just passed the one billion mark and is expected to double by 2050 (Branca et al. 2012). It is growing at 1.2 to 1.4% per year in South Africa and sub-Saharan Africa as a whole (Thornton 2010; Statistics South Africa 2017). Since the middle of the twentieth century, global agricultural output has kept pace with the rapidly growing population (Burney et al. 2010), with agriculture being the primary sector of most African countries (Saghir 2014).

South African agriculture is dualistic in nature, consisting of the less developed smallholder and well-developed commercial sectors (Vink and Kirsten 2003; DEA 2016). Commercial agricultural activities range from the intensive production of vegetables, ornamentals and other niche products to large-scale production of annual cereals, oilseeds, perennial herbaceous crops and tropical, subtropical and temperate fruit crops (DEA 2016). There are over 3 million smallholder farmers and 30,000 commercial farmers in South Africa (Armour 2014). Cultivated soils are generally very low in organic matter and are susceptible to wind erosion (FAO 2005; Du Preez et al. 2010).

5.2 Agricultural Land Use and Cropping Systems in South Africa

5.2.1 Agricultural Land in South Africa

More than 80% of South Africa's land is classified as either semiarid or arid, and 18% is dry sub-humid (FAO 2005). About 80% of total land in the country is used for agricultural purposes (DAFF 2015; FAOSTAT 2018), and only 14% of the agricultural land is arable (Fig. 5.1) and receives sufficient rainfall for crop production (FAO 2005; DAFF 2015). It was estimated that 12.2% of the land in the country was under cultivation in the year 2000 (FAO 2005). Croplands include annual commercial crops, annual semi-commercial or subsistence crops (DEA 2014). Field crops occupy 92% of the arable land and maize alone accounts for 51% of the land (FAO 2005). Over 90% of the cropland in the country is used to produce cereals. Perennial crops (orchards, viticulture and sugarcane) contribute about 8% towards the cropland area (DEA 2014). In South Africa, maize is the most important grain crop, being both the major feed grain and the national staple food (DAFF 2013). The total area planted to deciduous fruit amounts to 74,246 hectares (NAMC 2007). The forestry sector is well regulated in the country (Blanchard et al. 2011).

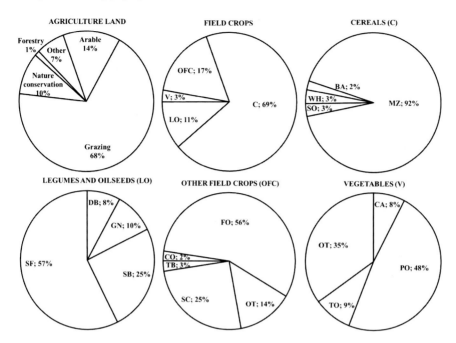

Fig. 5.1 Proportions of agricultural lands (DAFF 2015) and field croplands (Tongwane et al. 2016) in South Africa. Field crops: Cereals (C) [BA—barley; MZ—maize; SO—sorghum; WH—wheat]; legumes and oilseeds (LO) [DB—dry bean; GN—groundnuts; SB—soya bean; SF—sunflower]; other field crops (OFC) [CO—cotton; FO—fodder; OT—other; SC—sugarcane; TB—tobacco]; vegetables (V) [CA—cabbage; OT—other; PO—potato; TO—tomato]

Croplands are increasing with time in South Africa due to conversions from other land types (DEA 2014). Cropland and grasslands are estimated to have increased by 16.7% and 1.2%, respectively, between 2000 and 2010 (DEA 2014). There was a 1.2% increase in the transformed land between 1994 and 2005 (Dippenaar-Schoeman et al. 2013). According to the national land-cover data, there was only a 2.8% increase in the land used to cultivate subsistence crops between 1990 and 2013/14 (DEA 2016). During the same period, there was a further 10.6% and 16.2% increase in the land used to produce commercial permanent orchards and vines respectively. Furthermore, there was a 220.2% increase and a 7.6% decrease in the land cultivated to classes of pivot annual commercial crops and non-pivot commercial crops respectively during the same period (DEA 2016). The main drivers of land use change include environmental, political and socioeconomic challenges (DEA 2016).

5.2.2 Manure Management in Cropping Systems in South Africa

Various types of synthetic fertiliser are used during crop production in South Africa (Tongwane et al. 2016). A generic total amount of synthetic nitrogen (N) used in the country in 2012 was 0.4 million tonnes (Mt) (Tongwane et al. 2016). The application of synthetic nitrogen N fertiliser, lime application rates and area planted varied per crop (FAO 2005; Tongwane et al. 2016; Du Plessis 2003). Maize crops and sugarcane consume the highest N fertiliser and lime due to their respective large planted areas (Table 5.1). Consumption of synthetic fertiliser grew from 0.2 Mt in 1955 to 1.2 Mt in 1981 (Vermeulen et al. 2012). Various types of synthetic N fertiliser that include ammonium and nitrate concentrates are used during crop production (Tongwane et al. 2016). Maize, sugarcane and fruits account for 62%, 9% and 7.6% of national N consumption, respectively, and vegetables and wheat use 5.1% each (DEA 2016; IFA 2013). Generally, maize accounts for between 40 and 49% of the total use of fertiliser (FAO 2005; IFA 2013; Smale et al. 2011). Sugarcane accounted for 18% of fertiliser use, the second highest after maize, and contributes 10% to the total value of production (FAO 2005). The horticultural and fruit crop sectors account for 20% of fertiliser consumption (FAO 2005). Burning in sugarcane is practiced to facilitate stalk harvest and transportation, and this practice is the main source of non-CO_2 emissions (Galdos et al. 2009; Thompson 2012). Very little work has been done in South Africa to quantify GHG emissions from burnt and trashed sugarcane systems (Eustice et al. 2011).

5.3 Crop Management Practices and Greenhouse Gas Emissions in South Africa

5.3.1 Total GHG Emissions

Total GHG emissions from crop management increased from 24.3 Mt CO_2 equivalent (CO_2e) in 2000 to 28.3 CO_2e in 2010. Emissions of N_2O from the application of synthetic N fertiliser to agricultural soils in the country are the largest source of the emissions accounting for 90% of the total emissions in 2000 and 73% in 2010. A total of 81% of the synthetic N fertiliser emissions is N_2O, while CO_2 from urea accounts for 19% of the emissions from this agricultural input (Tongwane et al. 2016). Although overall emissions grew by 1.5% per annum, the largest increase by GHG type was CO_2 with 40% per annum. Total CH_4 and N_2O emissions increased and decreased at an annual rate of 0.4% and 0.6%, respectively. The emissions from soil management between 2000 and 2010 increased from 14.9 and 17.8 Mt CO_2e in 1990 and 1994, respectively (Blignaut et al. 2005). Production of field crops alone emitted a national total of 5.2 Mt CO_2e emissions in 2012 which account for approximately 17% of average annual emissions from agriculture,

Table 5.1 Average synthetic N fertiliser and lime application rates in South Africa, and area planted to selected main crops between 2000 and 2010

Crops	N fertiliser rate (kg/ha)	Lime (tonnes/ha) [after years][b]	2000	2001	2002	2003	2004	2005	2006	2007	2008	2009	2010
			Area planted ($\times 10^3$ ha)[d]										
Maize	55[a], 52.8[b], 40[c], 30[c], 20[c]	2.0 [4.0]	3189	3533	3651	3204	3223	2032	2897	3297	2896	3263	2859
Wheat	30[a,b]	1.9 [3.0]	934	974	941	748	830	805	765	632	748	642.5	558.1
Sunflower	15[a,b]	1.7 [3.0]	522	668	606	530	460	472	316	564	636	635.8	397.7
Sorghum	30[b]	2.3 [2.0]	88	75	95	130	86	37	69	87	86	87	69
Dry bean	24.9[b]	2.4 [4.0]	78	45	51	56	49	55	51	44	44	44	42
Soya beans	7[a], 19[b]	2.7 [3.0]	134	124	100	135	150	241	183	165	238	237.8	311.4
Sugarcane	92[a], 76[b]	3.3 [3.0]	429	432	430	427	425	428	420	423	389	382	376
Barley			78	73	72	84	83	90	90	73	68	75	83
Tobacco	38[a]	2.0 [3.0]	15	14.7	13.6	11.5	9.2	6	6	3.4	3.6	4	5.4
Cotton	36[a], 43.6[b]	2.0 [2.0]	57	39	23	36	22	18	11	9	6.8	5.1	13.1
Groundnuts	180[a,b]	2.0 [5.0]	165	94	50	72	40	49	41	54	55	57.4	55.1
Canola			19	27	31	44	44	40	32	33	34	35	34.8

[a]FAO (2005)
[b]Tongwane et al. (2016)
[c]Du Plessis (2003). Values depend on row width
[d]DAFF (2012)

forestry and other land use (Tongwane et al. 2016; DEA 2013). Maize, wheat and sugarcane are the main producers of the emissions by land area. Retaining of crop residues in the field after harvest accounted for 13% of the total national emissions from field crops (Tongwane et al. 2016).

5.3.2 Total CO_2 Emissions

Despite the general increase of GHG emissions, croplands were net sinks between 2003 and 2005 (DEA 2014, 2016). Croplands varied from a weak sink of 0.5 Mt CO_2 and a source of 7.5 Mt CO_2 between 2000 and 2010 (Fig. 5.2) (DEA 2014). Land conversions to croplands during the 2005–2010 period were responsible for the increased CO_2 emissions (DEA 2014). Land conversions that occurred in 2005 created a CO_2 source of 0.7 Mt in 2006 (DEA 2014). The release of N by mineralisation of soil organic matter as a result of change of land use or management contributes to an additional source of emissions (IPCC 2006). Cropland accounts for less than 1% and 21% of the national total GHG emissions and agricultural emissions, respectively (DEA 2014). The previous GHG emission report (DEA 2011) had estimated that cropland was a sink of 7 Mt in 2000. The emissions from croplands have high uncertainty levels as a result of constraints to activity data being publicly accessible and slightly different categories that are used by different data sources to classify this land use (DEA 2011, 2014, 2016; Stevens et al. 2016). More detailed cropland data that includes pivot and non-pivot systems needs to be used during estimations of the emissions (Stevens et al. 2016).

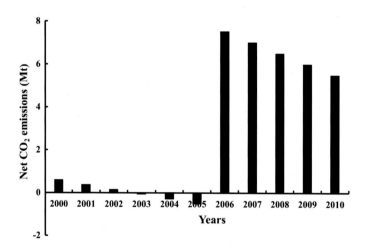

Fig. 5.2 Net CO_2 emissions from croplands in South Africa between 2000 and 2010. *Data source* DEA (2016)

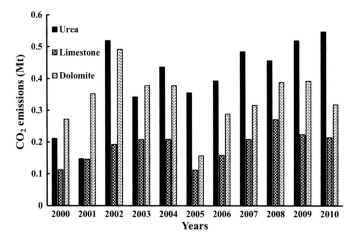

Fig. 5.3 CO_2 emissions from urea, limestone and dolomite in South Africa between 2000 and 2010. *Data source* DEA (2014)

Emissions from urea increased by 14.4% per annum from 0.2 Mt CO_2 in 2000 to 0.5 Mt CO_2 in 2010. Increases in market prices of urea post-2000 were lower than the prices of other types of synthetic fertilisers in the country (Grain SA 2011) and that could have enhanced its overall consumption and ultimate emissions. However, emissions from urea may be overestimated due to high uncertainties in the urea amounts used in the fields (DEA 2014, 2016; Grain SA 2011). Emissions from limestone and dolomite increased by annual rates of 8.4% and 1.5% between 2000 and 2010, respectively (DEA 2014) (Fig. 5.3). CO_2 from liming shows high annual variability that may be influenced by seasonal rainfall, but the general trend shows a slow increase of emissions from this agricultural activity (DEA 2014, 2016; DAFF 2010). Contribution of lime to total CO_2 emissions in 2010 is lower than in 2000 (Fig. 5.4) due to increases of emissions from net cropland and urea. Annual CO_2 emissions from the application of lime in agricultural lands contribute an average total of 1.5 Mt per year (Tongwane et al. 2016). Cereal crops, especially maize, are the major sources of emissions from this agricultural input (Tongwane et al. 2016). Emissions from urea and lime are highest in the regions that predominantly produce cereals (Free State, Mpumalanga and North West provinces) due to increasing croplands in these areas (DEA 2016).

5.3.3 Aggregated Non-CO_2 Emissions

Non-CO_2 emissions contributed 95% of the total GHG emissions in 2000 and 77% in 2010. There is generally a slight decline of N_2O emissions from crop management in the country, from 22.0 Mt CO_2e in 2000 to 20.6 Mt CO_2e in 2010.

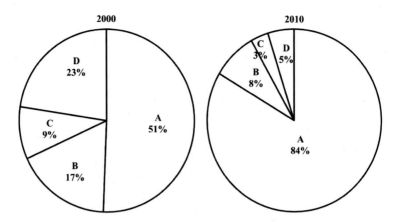

Fig. 5.4 Comparison of CO_2 sources in South Africa, 2000 and 2010 [A—cropland; B—urea; C—limestone; D—dolomite]

Applications of N fertiliser to soils are main sources of N_2O emissions (DEA 2014; Tongwane et al. 2016). Direct N_2O emissions fluctuated annually with the year 2000 having the highest and 2009 the lowest emissions of 16.1 Mt CO_2e and 14.6 Mt CO_2e, respectively (DEA 2014). The emissions of N_2O occur directly from the soils to which the N is added and through two indirect pathways (i.e. (i) volatilisation of NH_3 and NO_x from managed soils and from fossil fuel combustion and biomass burning and (ii) leaching and run-off of N from managed soils) (IPCC 2006).

Application of synthetic fertiliser to soils is the main source of GHG emissions from production of field crops in South Africa with a national total of 3.0 Mt of CO_2e (Fig. 5.5) (Tongwane et al. 2016). High emissions from synthetic fertiliser are

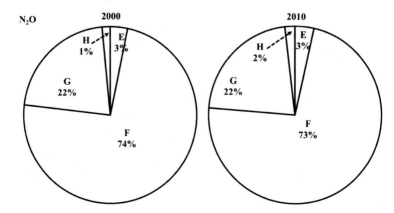

Fig. 5.5 Sources of N_2O emissions from crop production in South Africa, 2000 and 2010 [E—biomass burning; F—direct soil emissions; G—indirect soil emissions; H—manure management]

Table 5.2 Direct N_2O emissions (Mt) from synthetic N fertiliser in South Africa between 2000 and 2010

	2000	2001	2002	2003	2004	2005	2006	2007	2008	2009	2010
Maize	2.76	3.05	3.16	2.77	2.79	1.76	2.50	2.85	2.50	2.82	2.47
Wheat	0.81	0.84	0.81	0.65	0.72	0.70	0.66	0.55	0.65	0.56	0.48
Sunflower	0.45	0.58	0.52	0.46	0.40	0.41	0.27	0.49	0.55	0.55	0.34
Sorghum	0.08	0.06	0.08	0.11	0.07	0.03	0.06	0.08	0.07	0.08	0.06
Dry bean	0.07	0.04	0.04	0.05	0.04	0.05	0.04	0.04	0.04	0.04	0.04
Soya beans	0.12	0.11	0.09	0.12	0.13	0.21	0.16	0.14	0.21	0.21	0.27
Sugarcane	0.37	0.37	0.37	0.37	0.37	0.37	0.36	0.37	0.34	0.33	0.32
Barley	0.07	0.06	0.06	0.07	0.07	0.08	0.08	0.06	0.06	0.06	0.72
Tobacco	0.01	0.01	0.01	0.01	0.01	0.01	0.01	0.00	0.00	0.00	0.05
Cotton	0.05	0.03	0.02	0.03	0.02	0.02	0.01	0.01	0.01	0.00	0.01
Groundnuts	0.14	0.08	0.04	0.06	0.03	0.04	0.04	0.05	0.05	0.05	0.05
Canola	0.02	0.02	0.03	0.04	0.04	0.03	0.03	0.03	0.03	0.03	0.03
Total	4.93	5.27	5.24	4.73	4.69	3.69	4.22	4.65	4.50	4.73	4.84

as a result of high application rates that are aimed at improving soil fertility and crop productivity (Tongwane et al. 2016). However, emissions from synthetic N fertilisers (Table 5.2) have generally not increased since 2000 (DEA 2014) probably due to high costs of these inputs. The prices of N fertilisers are directly related to the price of natural gas which, on the other hand, is highly influenced by the crude oil prices (Grain SA 2011). Contributions to total national GHG emissions from field crops vary significantly between different crops (Tongwane et al. 2016).

Direct and indirect N_2O emissions from the application of synthetic N fertiliser on managed lands vary slightly from year to year, depending on the planted area (Tables 5.2 and 5.3). For field crops, maize production is the largest source of

Table 5.3 Indirect N_2O emissions (Mt) from synthetic N fertiliser in South Africa between 2000 and 2010

	2000	2001	2002	2003	2004	2005	2006	2007	2008	2009	2010
Maize	0.28	0.31	0.32	0.28	0.28	0.18	0.25	0.28	0.25	0.28	0.25
Wheat	0.08	0.08	0.08	0.06	0.07	0.07	0.07	0.05	0.06	0.06	0.05
Sunflower	0.05	0.06	0.05	0.05	0.04	0.04	0.03	0.05	0.05	0.05	0.03
Sorghum	0.01	0.01	0.01	0.01	0.01	0.00	0.01	0.01	0.01	0.01	0.01
Dry bean	0.01	0.00	0.00	0.00	0.00	0.00	0.00	0.00	0.00	0.00	0.00
Soya beans	0.01	0.01	0.01	0.01	0.01	0.02	0.02	0.01	0.02	0.02	0.03
Sugarcane	0.04	0.04	0.04	0.04	0.04	0.04	0.04	0.04	0.03	0.03	0.03
Barley	0.01	0.01	0.01	0.01	0.01	0.01	0.01	0.01	0.01	0.01	0.07
Tobacco	0.00	0.00	0.00	0.00	0.00	0.00	0.00	0.00	0.00	0.00	0.00
Cotton	0.00	0.00	0.00	0.00	0.00	0.00	0.00	0.00	0.00	0.00	0.00
Groundnuts	0.01	0.01	0.00	0.01	0.00	0.00	0.00	0.00	0.00	0.00	0.00
Canola	0.00	0.00	0.00	0.00	0.00	0.00	0.00	0.00	0.00	0.00	0.00
Total	0.49	0.53	0.52	0.47	0.47	0.37	0.42	0.47	0.45	0.47	0.48

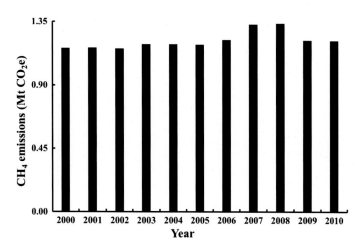

Fig. 5.6 CH$_4$ emissions from burning of agricultural biomass between 2000 and 2010. *Data source* DEA (2016)

emissions (Tongwane et al. 2016). However, due to a general decrease of N fertiliser applied to managed lands (DEA 2014), there is a reduction of emissions from this input. The production of maize has been declining over the last two decades, and with maize being the biggest consumer of N fertiliser, the emissions from this input have slowed (DEA 2014). However, due to increasing production area of soya beans in the country (DAFF 2015), emissions from this crop commodity show a rapid increase. Synthetic N fertiliser contributes more than half of the total emissions from field crops (Tongwane et al. 2016).

Burning of agricultural biomass resulted in annual CH$_4$ emissions of approximately 1.2 Mt CO$_2$e per year between 2000 and 2006 (Fig. 5.6) (DEA 2014). The emissions increased and peaked in 2007 and 2008 with 1.3 Mt CO$_2$e as a result of increased average percentage of area burnt (DEA 2014). The burning of biomass is classified into the six land use categories defined in the IPCC guidelines (forest, grassland, cropland, wetlands, settlements and other land) (DEA 2014). It is estimated that croplands contribute 14.6% of emissions from biomass burning.

Total N$_2$O emissions from management of crop residues are gradually decreasing over time (Table 5.4). There is a decline in emissions despite a general increase of emissions from management of maize residues. Main decreases of emissions are from wheat and sugarcane as a result of reduction of production areas of these crops over time. On the other hand, emissions from maize residues increased due to improved yields. The share of maize residues to the total emissions from crop residues increased from 22.0% in 2000 to 38.0% in 2010, while the contributions of wheat and sugarcane decreased from 9.0% to 7.0% and 61.0% to 49.0%, respectively, during the period. The emissions from residues of sunflower and sorghum are variable with time and do not show a clear trend with regard to their overall share to the total emissions from management of crop residues.

Table 5.4 N$_2$O emissions (Mt) from management of crop residues in South Africa between 2000 and 2010

	2000	2001	2002	2003	2004	2005	2006	2007	2008	2009	2010
Maize	0.362	0.469	0.451	0.453	0.547	0.323	0.341	0.612	0.585	0.624	0.508
Wheat	0.149	0.153	0.149	0.095	0.103	0.117	0.129	0.117	0.132	0.120	0.088
Sunflower	0.044	0.062	0.044	0.043	0.041	0.035	0.020	0.058	0.053	0.033	0.057
Sorghum	0.009	0.011	0.011	0.019	0.013	0.005	0.008	0.012	0.013	0.009	0.007
Dry bean	0.007	0.004	0.005	0.006	0.005	0.005	0.003	0.004	0.005	0.004	0.003
Soya beans	0.000	0.000	0.000	0.000	0.000	0.000	0.000	0.000	0.000	0.000	0.000
Sugarcane	0.984	0.872	0.948	0.841	0.787	0.867	0.836	0.813	0.793	0.769	0.660
Barley	0.004	0.005	0.007	0.009	0.007	0.009	0.009	0.009	0.009	0.008	0.007
Tobacco	0.029	0.027	0.031	0.021	0.019	0.012	0.010	0.007	0.008	0.010	0.001
Cotton	0.020	0.011	0.010	0.017	0.013	0.009	0.007	0.006	0.005	0.005	0.010
Groundnuts	0.017	0.010	0.005	0.010	0.006	0.007	0.005	0.008	0.009	0.008	0.006
Canola	0.001	0.002	0.002	0.003	0.002	0.003	0.002	0.003	0.002	0.003	0.003
Total	1.626	1.626	1.663	1.516	1.544	1.391	1.371	1.649	1.614	1.593	1.351

5.4 Mitigation of GHG Emissions from Cropping Systems in South Africa

5.4.1 Agricultural Baseline Emissions

Agricultural baseline emissions are predicted to increase from 50.6 Mt CO$_2$e in 2010 to 69.6 Mt CO$_2$e in 2050 (DEA 2016; Stevens et al. 2016). Projections indicate that the area planted to yellow maize will exceed that planted to white maize in the near future given current consumption patterns that result in a flat demand for white maize in the food consumption market, compared to the continued growth in demand for animal feed (DEA 2016). This suggests that fertilisation rates will increase accordingly. Future baseline estimates show a gradual linear increase of emissions from synthetic N fertiliser and lime (Fig. 5.7). Emissions from urea are projected to increase exponentially. The largest increase in the baseline emissions comes from emissions from urea application; however, the urea consumption data is highly variable and comes with high uncertainties as it

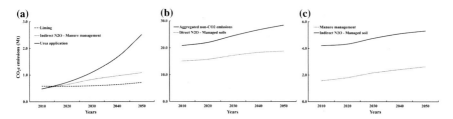

Fig. 5.7 Baseline emissions for the agricultural sector in South Africa between 2010 and 2050. *Data source* Stevens et al. (2016)

determined from import and export data with the assumption that all urea is being applied to the field (DEA 2016). Because of demand for human food and animal feed, field crops are expected to be the main sources of these emissions.

Direct N_2O emissions from managed lands are the largest contributor to the baseline emissions from aggregated and non-CO_2 emissions (DEA 2016). In terms of reducing non-CO_2 emissions from management of soils, options available to the country include improved fertiliser usage and an increase in production of legumes (DEA 2016). The principal biological N-fixing crops are soya beans, groundnuts and lucerne (DEA 2014). Mitigation strategies to lower the N_2O emissions from the agricultural sources should be put in place and be promoted among farmers. In order to reduce emissions, the improvements in agricultural technologies and practices need to achieve sustainable agricultural production, accrual of additional benefits to the farmer and agricultural products that are accepted by consumers (IPCC 1996).

5.4.2 Biofuels

In Africa today, as in most parts of the world, the biofuel industry is receiving serious attention in view of the vast land available and the favourable climate for growing many of the energy crops (Marvey 2009). Biofuels are among the highest renewable energy sources in South Africa, with an estimated contribution of 9.0–14.0% (Sawahel 2016). South Africa was the first southern African country to develop a formal biofuel strategy (Blanchard et al. 2011). The National Biofuels Industrial Strategy focused on a 5-year pilot programme to achieve a 2.0% penetration of biofuels in the national liquid fuel supply or 400 million litres per year—to be based on local agricultural and manufacturing production capacity (Blanchard et al. 2011; DME 2007). The strategy aimed to achieve economic and social development in rural areas via the agricultural development in the former homeland areas (Blanchard et al. 2011). Canola, sunflower and soya beans are promoted as feedstocks for biodiesel, while sugarcane and sugar beet are the choice of feedstocks for bioethanol (Marvey 2009; Pradhan and Mbohwa 2014). However, with the exception of canola which is a dry land winter crop suitable as a rotational and complementary crop for existing oilseed crops, the production figures of oilseeds indicate a general decline in yields and a corresponding decrease of the area planted (Marvey 2009). Maize is excluded from these feedstocks because its use may compromise national food security (Marvey 2009; Pradhan and Mbohwa 2014). Several companies showed interest to invest in biofuel projects (Marvey 2009). However, progress in the development of the country's biofuel industry remains slow at present (Van Zyl and Prior 2009). The South African strategy is generally considered to be conservative, tempering the international drive towards large-scale biofuel production with a pragmatic approach (WRC 2009). As a result, the South African government faces the challenge of showing a strong commitment to the biofuel industry through their policy regulations and incentives (Marvey 2009).

Incentives were only provided for locally based processing plants that relied on feedstocks being acquired via contractual agreements from small-scale farmers (Blanchard et al. 2011). The combination of the preference given to previously disadvantaged farmers and the exclusion of maize as feedstock has slowed down the establishment of an agriculture-based biofuels industry in South Africa (Letete and Von Blottnitz 2012).

Biofuel developments are still at an early phase, and ongoing research to optimise feedstocks and processing techniques may well promote feedstocks not mentioned in the strategy (Blanchard et al. 2011). The ability of biofuel crops to mitigate GHG emissions varies widely between crops, management practices and the nature of the land where the biofuel crop is grown (Von Maltitz 2017). About 30.0% and 50.0% reduction in GHG emissions can be achieved from ethanol and biodiesel, respectively (DOE 2013). Biofuels may have a restorative capability, increasing soil productivity and biodiversity within an agro-ecological system (Blanchard et al. 2011).

5.5 Conclusions

Crop management practices affect GHG emissions in South Africa. Application of synthetic fertiliser to the soil resulted in the highest GHG emissions with 57% of the total national emissions from production of field crops in the country. Application of lime during production of field crops and crop residues retained in the field after harvest resulted in 30.0% and 13.0% of the total emissions from field crops, respectively. Cereal crops are responsible for 68% of the national total emissions with maize contributing 56.0% of the national total. Production of maize, wheat and sugarcane resulted in the highest commodity GHG emissions in the country in 2012. Crop management practices that include use of improved technologies and fertilisation rates have a considerable effect on the amount of GHG emissions from crop production. However, agricultural croplands that are intensively managed offer many opportunities for reducing GHG emissions through changes in agronomic practices.

References

Armour J (2014) Dualism in SA agriculture. In: Proceedings of the FERTASA 54th annual congress, 10 June 2014, Johannesburg, South Africa

Blanchard R, Richardson DM, O'Farrell PJ, Von Maltitz GP (2011) Biofuels and biodiversity in South Africa. S Afr J Sci 107(5/6), Article no. 186

Blignaut JN, Chitiga-Mabugu MR, Mabugu RM (2005) Constructing a greenhouse gas emissions inventory using energy balances: the case of South Africa for 1998. J Energy S Afr 16(3):21–32

Branca G, Tennigkei T, Mann W, Lipper L (2012) Identifying opportunities for climate-smart agriculture investments in Africa. Food and Agriculture Organization of the United Nations, Rome

Burney JA, Davis SJ, Lobell DB (2010) Greenhouse gas mitigation by agricultural intensification. PNAS 107(26):12052–12057

DAFF (2010) Quarterly economic review of the agriculture sector. Department of Agriculture, Forestry and Fisheries, Pretoria, South Africa

DAFF (2012) Abstract of agricultural statistics 2012. Department of Agriculture, Forestry and Fisheries, Pretoria, South Africa

DAFF (2013) Trends in the agricultural sector 2013. Department of Agriculture, Forestry and Fisheries, Pretoria, South Africa

DAFF (2015) Abstract of agricultural statistics 2015. Department of Agriculture, Forestry and Fisheries, Pretoria, South Africa

DEA (2011) South Africa's second national communication under the united nations framework convention on climate change. Department of Environmental Affairs, Pretoria, South Africa

DEA (2013) Long term adaptation scenarios, agriculture and forestry. Department of Environmental Affairs, Pretoria, South Africa

DEA (2014) GHG inventory for South Africa: 2000–2010. Department of Environmental Affairs, Pretoria, South Africa

DEA (2016) Towards the development of a GHG emissions baseline for the agriculture, forestry and other land use (AFOLU) sector in South Africa (2016). Department of Environmental Affairs, Pretoria, South Africa

Dippenaar-Schoeman AS, Van den Berg AM, Lyle R, Haddad CR, Foord SH, Lotz LN (2013) Die diversiteit van Suid-Afrikaanse spinnekoppe (Arachnida: Araneae): dokumentering van 'n nasionale opname [The diversity of South African spiders (Arachnida: Araneae): Documenting a national survey]. Tydskrif van die Suid- Afrikaanse Akademie vir Wetenskap en Kuns 32 (375):1–7

DME (2007) Biofuels industrial strategy of the republic of South Africa. Department of Minerals and Energy, Pretoria, South Africa

DOE (2013) Mandatory blending of biofuels with petrol and diesel to be effective from the 01 October 2015. Media Statement, Department of Energy, Pretoria, South Africa

Du Plessis J (2003) Maize production. Department of Agriculture, Pretoria, South Africa

Du Preez C, Mnkeni P, Van Huyssteen C (2010) Knowledge review on land use and soil organic matter in South Africa. In: Proceedings of the 19th world congress of soil science, soil solutions for a changing world, 1–6 August 2010, Brisbane, Australia

Eustice T, Van Der Laan M, Van Antwerpen R (2011) Comparison of greenhouse gas emissions from trashed and burnt sugarcane cropping systems in South Africa. Proc S Afr Sugar Technol Assoc 84:326–339

FAO (2005) Fertilizer use by crop in South Africa. Land and Plant Nutrition Management Service, Land and Water Development Division, Food and Agriculture Organization of the United Nations, Rome. http://www.fao.org/docrep/008/y5998e/y5998e00.HTM

FAO (2014) FAOSTAT: Emissions—land use. http://faostat3.fao.org/faostat-gateway/go/to/download/G2/*/E. Accessed 22 June 2018

FAOSTAT (2018) Resources. Food and Agricultural Organization of the United Nations Statistics. Rome. http://faostat.fao.org/site/377/default.aspx#ancor. Accessed 26 June 2018

Galdos MV, Cerri CC, Cerri CEP (2009) Soil carbon stocks under burned and unburned sugarcane in Brazil. Geoderma 153:347–352

Grain SA (2011) Grain SA fertiliser report. Pretoria, South Africa

IFA (2013) Assessment of fertilizer use by crop at the global level 2010–2010/11. International Fertilizer Industry Association, Paris, France

IPCC (Intergovernmental Panel on Climate Change) (1996) Greenhouse Gas Inventory Reporting Instructions. Revised 1996 IPCC Guidelines for National Greenhouse Gas Inventories. UNEP, WMO, OECD and IEA, Bracknell, UK

IPCC (Intergovernmental Panel on Climate Change) (2006) IPCC guidelines for national greenhouse gas inventories. National Greenhouse Gas Inventories Programme, Japan

IPCC (Intergovernmental Panel on Climate Change) (2014) Climate change 2014: mitigation of climate change. In: Edenhofer O, Pichs-Madruga R, Sokona Y, Farahani E, Kadner S, Seyboth K, Adler A, Baum I, Brunner S, Eickemeier P, Kriemann B, Savolainen J, Schlömer S, von Stechow C, Zwickel T, Minx JC (eds) Contribution of working group III to the fifth assessment report of the intergovernmental panel on climate change. Cambridge University Press, Cambridge, United Kingdom and New York, NY, USA

Letete TCM, Von Blottnitz H (2012) Biofuel policies in South Africa: a critical analysis. In: Janssen R, Rutz D (eds) Bioenergy for sustainable development in Africa. Springer, South Africa

Marvey BB (2009) Oil crops in biofuel applications: South Africa gearing up for a bio-based economy. TD J Transdiscipl Res South Africa 5(2):153–161

NAMC (2007) Subsector study: deciduous fruit. The national Agricultural Marketing Council and Commark Trust. Report No 2007-02. Pretoria, South Africa

Pradhan A, Mbohwa C (2014) Development of biofuels in South Africa: challenges and opportunities. Renew Sustain Energy Rev 39:1089–1100

Saghir J (2014) Global challenges in agriculture and the World Bank's response in Africa. J Food and Energy Secur 3(2):61–68

Sawahel W (2016) Brazil and India join senegal for biofuel production. http://www.scidev.net/global/biofuels/news/brazil-and-india-join-senegal-for-biofuel-producti.html. Accessed 18 June 2018

Smale DA, Wernberg T, Peck LS, Barnes DKA (2011) Turning on the heat: ecological response to simulated warming in the sea. PLoS ONE 6(1):e16050

Statistics South Africa (2017) Mid-year population estimates, 2017. Statistics South Africa, Pretoria. http://www.statssa.gov.za/publications/P0302/P03022017.pdf. Accessed 05 July 2018

Stevens LB, Henri AJ, van Nierop M, van Starden E, Lodder J, Piketh S (2016) Towards the development of a GHG emissions baseline for the agriculture, forestry and other land use (AFOLU) sector, South Africa. Clean Air J 26(2):34–39

Thompson P (2012) The agricultural ethics of biofuels: the food vs. fuel debate. Agriculture 2(4):339–358

Thornton PK (2010) Livestock production: recent trends, future prospects. Philos Trans R Soc 365:2853–2867

Tongwane M, Mdlambuzi T, Moeletsi M, Tsubo M, Mliswa V, Grootboom L (2016) Greenhouse gas emissions from different crop production and management practices in South Africa. Environ Develop 19:23–35

Van Zyl WH, Prior BA (2009) South Africa biofuels. IEA Task Group Progress Report 39. Biofuels, Stellenbosch, South Africa. http://academic.sun.ac.za/biofuels/media%20info/South%20Africa%20Biofuels%20May%202009%20Progress%20Report.pdf. Accessed 27 June 2018

Vermeulen SJ, Campbell B, Ingram J (2012) Climate change and food systems. Annu Rev Environ Resour 37:195–222

Vink N, Kirsten J (2003) Agriculture in the national economy. In: Nieuwoudt L, Groenewald J (eds) The challenge of change: agriculture, land and the South African economy. University of Natal Press, Pietermaritzburg

Von Maltitz G (2017) Options for suitable biofuel farming: experience from Southern Africa, WIDER Working Paper, No. 2017/100. ISBN 978-92-9256-324-0, UNU-WIDER, Helsinki

WRC (2009) Scoping study on water use of crops/trees for biofuels in South Africa. Water Research Commission Report No. 1772/1/09. Pretoria, South Africa

Chapter 6
Agricultural Greenhouse Gases from Sub-Saharan Africa

Kofi K. Boateng, George Y. Obeng and Ebenezer Mensah

Abstract Climate change has variously been diagnosed as perhaps the most challenging issue that confronts the twenty-first century, and especially for sub-Saharan Africa (SSA), the impacts of a changing climate have already been felt in most regions and in various sectors of the economy principally, agriculture. Agriculture on the subcontinent, although still very rudimentary in terms of management practices and production efficiency, provides the mainstay for majority of the people and is heavily climate dependent. This makes climate change an issue requiring immediate and effective interventions, viz. adaptation and resilience building to safeguard the livelihood of over a billion people. This chapter looks at sub-Saharan African agriculture, its contribution to the emission of greenhouse gases and their pathways by using the FAOSTAT system and the other literature on emission research from peer-reviewed journals. An attempt is also made to gauge the effects of a changing climate on SSA agricultural productivity. The contribution of SSA agriculture to the socio-economic well-being of its people is also discussed. Adaptation and resilience building among the dominating smallholder farmers in the region are captured, and the factors that hinder the effective scaling up of strategies aimed at ameliorating the effects of climate variability on local agriculture.

Keywords Sub-Saharan Africa agriculture · Climate change · Greenhouse gas emissions

---

K. K. Boateng (✉) · E. Mensah
Department of Agricultural and Biosystems Engineering, Kwame Nkrumah
University of Science and Technology, UPO, KNUST, Kumasi, Ghana
e-mail: edkoboat@hotmail.com

E. Mensah
e-mail: ebenmensah@gmail.com

G. Y. Obeng
Technology Consultancy Center and Mechanical Engineering Department,
College of Engineering, Kwame Nkrumah University of Science and Technology,
UPO, KNUST, Kumasi, Ghana
e-mail: george.yaw.obeng@asu.edu; geo_yaw@yahoo.com

© Springer Nature Singapore Pte Ltd. 2019
N. Shurpali et al. (eds.), *Greenhouse Gas Emissions*, Energy, Environment,
and Sustainability, https://doi.org/10.1007/978-981-13-3272-2_6

6.1 Sub-Saharan African Agriculture

The FAO puts agriculture in sub-Saharan Africa (SSA) as the major source of livelihood for a population of approximately 1 billion people (Alexandratos and Bruinsma 2012) as it contributes immensely to the economies of most economies in the region (Jerven and Duncan 2012).

Although agricultural contribution to GDP has generally declined over the years in most sub-Saharan African economies as a result of economic diversification, the agricultural sector still remains the major employer in most of these countries. According to the International Monetary Fund (IMF), agriculture in SSA is responsible for the direct employment of half of the sub-region's labour force. For the rural populations, it remains the mainstay for a significant proportion of people as smallholder farmers form 80% of production and this is estimated to give direct employment to approximately 176 million people on the subcontinent (IMF 2012).

The agriculture sector in terms of total employment employs 42% of Ghanaians, 28% of Nigerians, 62% of Kenyans, 75% of Mozambicans and Rwandese and 68% of Zimbabweans. Overall, the agricultural sector employs 55% of sub-Saharan Africans (Schlenker and Lobell 2010a). Contrary to the importance of the agricultural sector to sub-Saharan African countries, the sector is the least developed in terms of infrastructure and production levels. Production remains at the smallholder level in most SSA countries with the use of crude farming tools still very dominant. Value addition to primary farm products to semi-finished and finished products remains a challenge to the sector resulting in huge post-harvest losses. Food losses equated to post-harvest losses alone in sub-Saharan Africa exceeded 30% of total crop production equivalent to USD $4 billion annually (OECD/FAO 2016) (Fig. 6.1).

Irrigation, considered as an important factor for agricultural growth in low-income countries, is insufficient in most parts of SSA making agriculture extensively rain-fed (Müller et al. 2011). Generally, irrigated area in sub-Saharan Africa makes up just 5% of its total cultivated area and two-thirds of this area is in three countries Madagascar, South Africa and Sudan. Thus, SSA lags significantly

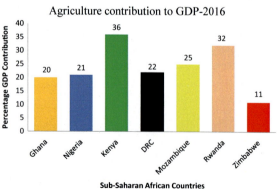

Fig. 6.1 Agricultural contribution to GDP (OECD/FAO 2016)

behind Asia and Latin America with 37% and 14% of their cultivated area under irrigation, respectively (You 2008).

In Ethiopia, irrigated agriculture constitutes only 1.1% of the total cultivated land (Bewket and Conway 2007) and less than 3% of the current food production in Ethiopia (Awulachew et al. 2005). In Ghana, of the 14,038,224 hectares of total agricultural land, 30,345 ha representing 0.4% is under irrigation (Ministry of Food and Agriculture 2013).

Overall, agriculture in SSA is predominantly rain-fed at 96% of overall crop production making agricultural production in sub-Saharan Africa particularly vulnerable to the effects of climate change (World Bank 2015; Yéo et al. 2016).

6.1.1 Climate Change Effects on Sub-Saharan African Agriculture

The Intergovernmental Panel on Climate Change (IPCC) identifies Africa as continent most vulnerable to the impacts of climate change (IPCC 2014). Projections that have been made for SSA point to an increasingly warming trend in the form of frequent occurrence of extreme heat events, increasing aridity and changes in rainfall patterns (Serdeczny 2016).

Climate change projections for SSA indicate a warming trend which will significantly distress natural and human systems, especially in the inland tropics where frequent occurrence of extreme heat events, changes in rainfall patterns, increasing aridity are expected to be pronounced (Serdeczny 2016). Agriculture in SSA is at significant risk under changing climate primarily due to its overdependence on rain as well as observed crop sensitivities to high temperatures during the growing season (Schlenker and Roberts 2009; Lobell et al. 2011). The lack of adaptive capacity, small farm sizes, low capitalization and limited use of improved technologies lowers the resilience and increases the vulnerability of smallholder farmers in the sub-region to the negative effects of climate change (Morton 2007:6).

Important crops grown in SSA in terms of the provision of calories, protein and fat for a significant percentage of the population are maize, cassava, rice, sorghum, wheat and millet in the order of their importance (FAO 2009). However, the most important crops grown in SSA in terms of area harvested are millet, maize, sorghum and cassava, cultivated on 50% of total harvested land (Blanc 2011). There is a high possibility that the total effect on yields from major crops in SSA due to climate change will be negative and devastating (Niang et al. 2014). In a study that modelled the impacts of climate change on sub-Saharan agriculture (Schlenker and Lobell 2010b), a negative growth to the extent—of 22, 17, 17 and 8% in the production of maize, sorghum, millet and cassava, respectively, has been projected by the middle of this century.

Overall, Africa is expected to experience mainly negative climate change impacts, in terms of an increase in the already high temperatures and a decrease in the largely erratic rainfall in its context of widespread poverty and low development (Speranza 2010). It is, therefore, important that appropriate climate adaptation and resilience building strategies are developed and effectively implemented so as not to exacerbate SSA climate risks.

6.1.2 Sources and Contribution of GHG Emissions from SSA Agriculture

On a global scale, GHG emissions from SSA agriculture are significant. The sector is the largest emitter of GHGs and currently accounts for 27% of the total emissions from the whole continent. The biggest emission sources in SSA agricultural sector include the conversion of forest to cropland and pasture, livestock manure and digestive processes, burning of savannah, cropland management and cultivation (management) practices (Hogarth et al. 2015).

This is evident from a review and synthesis of greenhouse gas emissions from 22 SSA countries (Kim 2015). GHG emission levels from natural and agricultural lands are shown in Table 6.2. Factors that were found to affect the emission levels included soil physical and chemical properties, rewetting, vegetation type, forest management and land-use changes (Table 6.1).

For levels of cropland GHG emissions, soil amendments with crop residues, inorganic fertilizers as well as manure are major determinants. Management practices employed by farmers, therefore, become critical in any emission reduction strategy.

East Africa has the highest level of emissions due to agricultural production for both methane and nitrous oxide. Current emission data indicates that the high level

Table 6.1 SSA emissions and sources (Kim 2015)

GHG	Range		
Carbon dioxide (CO_2)	3.3–57.0 Mg ha^{-1} year^{-1}		
Methane (CH_4)	−4.8 to 3.5 kg ha^{-1} year^{-1}		
Nitrous oxide (N_2O)	−0.1 to 13.7 kg ha^{-1} year^{-1}		
General sources of GHG emissions			
	CO_2	CH_4	N_2O
Aquatic systems	5.7–232.0 Mg ha^{-1} year^{-1}	−26.3 to 2741.9 kg ha^{-1} year^{-1}	0.2–3.5 kg ha^{-1} year^{-1}
Croplands		−1.3 to 1566.7 kg ha^{-1} year^{-1}	0.05–112.0 kg ha^{-1} year^{-1}
Vegetable gardens	73.3–132.0 Mg ha^{-1} year^{-1}	–	53.4–177.6 kg ha^{-1} year^{-1}
Agroforestry	38.6 Mg ha^{-1} year^{-1}	–	0.2–26.7 kg ha^{-1} year^{-1}

6 Agricultural Greenhouse Gases from Sub-Saharan Africa

Table 6.2 Regional emission levels in GgCO$_2$e (Kim 2015)

Region	CH$_4$	N$_2$O	Total agricultural emissions
West Africa	116,959	93,600	210,560
East Africa	204,275	172,238	376,512
Central Africa	51,418	55,937	107,355
Southern Africa	22,984	24,268	47,251
Total	395,636	346,042	741,677

of emissions in the region is attributable to the high livestock activity (enteric fermentation) in the region.

6.1.3 Enteric Fermentation

Livestock rearing is an important aspect of agriculture in Africa with an estimated herd of 981 million cattle, goats and sheep (Hogarth et al. 2015). The herd size is projected to increase as a result of high demand for meat and milk owing to rapid population growth. Most of this growth in livestock population is expected to occur in East and Western Africa exacerbating their enteric emission footprints (Herrero et al. 2008) (Fig. 6.2).

6.1.4 Paddy Rice Cultivation

For the West African sub-region, crop cultivation is the major source of agricultural greenhouse gas emissions, especially paddy rice cultivation. Methane (CH$_4$) and

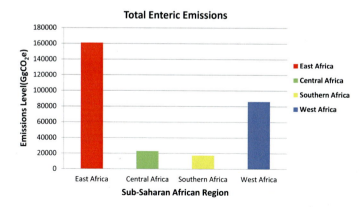

Fig. 6.2 2016 enteric fermentation (livestock emissions) (FAOSTAT 2016)

nitrous oxide (N_2O) emissions are enhanced in paddy rice cultivation through flooding and fertilizer application regimes, two key management practices farmers employ.

For most regions of SSA, rice remains a critical staple food for most households and its demand outstrips all other cereals except maize (Tollens 2006). The demand for rice on the subcontinent continues to outpace local production. The region, however, is only able to meet 50% of this growing demand. Paddy rice production in SSA with a total harvested area of 11,815,947 ha currently stands at 26,116,184 tons (FAOSTAT 2016). Western and Eastern Africa lead in production with 66% and 30%, respectively. The combined production total of Central and Southern Africa is less than half the production of East Africa, the second largest paddy rice production area on the subcontinent. For most countries in SSA, rice production has become an important sector with many countries having drawn up national rice development strategies to bolster production. West Africa currently leads in emissions from paddy rice production at 15991 $GgCO_2e$, and with rice production projected to grow to meet an increasing demand, it is expected that emissions from rice production will also increase (Fig. 6.3).

6.1.5 Emissions from Synthetic Fertilizer Use

The primary GHG that is emitted from synthetic fertilizer use is nitrous oxide, and it is produced by the microbial process of nitrification and denitrification. This process gives rise to direct N_2O emissions. Indirect N_2O emissions arise when volatilization and leaching processes commence (Linquist 2012; Liang 2013; Xia 2016). Agriculture is responsible for 85% of N_2O emissions globally (Syakila and Kroeze 2011; Signor et al. 2013).

The use of synthetic fertilizers is low in sub-Saharan Africa compared to other regions of the world. Currently, SSA agriculture consumes 15 kg of fertilizer per hectare of arable land (World Bank 2016). As a consequence, emissions from

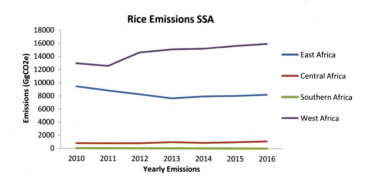

Fig. 6.3 SSA emission trends 2010–2016 (FAOSTAT 2016)

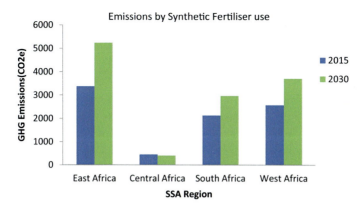

Fig. 6.4 Emissions by synthetic fertilizer use (FAOSTAT 2016)

synthetic fertilizer use are low in SSA. Future projections of emissions point to an average increase in emissions by 25% with East and West Africa emissions being the highest (FAOSTAT 2016) (Fig. 6.4).

Synthetic fertilizer use is projected to increase due to the pressure on food production to increase to meet the demand of a growing population in SSA. The traditional fallow periods required for the recovery of depleted soil nutrients is no longer a viable nutrient management option due to the long period it takes for soils to replenish lost nutrients. Synthetic fertilizer usage, therefore, becomes the best alternative to boost food crop production. The challenge, therefore, for SSA agriculture is to avoid the negative effects and the extensive use of synthetic fertilizer usage. High nutrient use efficiency has been advocated as an effective means of ensuring that SSA increases crop production with synthetic fertilizer use without maximizing its adverse effects including N_2O emissions. An approach to efficient fertilizer use through a 4R guide as described in the following has been proposed (Richards 2016):

1. Use the **right** source of nutrients (the right composition of nutrients, including other than NPK),
2. Applied at the **right** rate (based on economic criteria and soil fertility status),
3. Applied at the **right** time (relative to crop needs and weather),
4. Applied at the **right** place (targeting plant roots and minimizing losses).

It is essential to apply good agronomic practices alongside efficient use of nutrient input to achieve high nutrient use efficiency (Richards 2016). Examples of such agronomic practices include the use of improved, high yielding varieties that can adapt to the local environment, application and recycling of available organic matter (crop residues and farmyard manure), water harvesting and irrigation under drought stress conditions, and lime application on soils with acidity-related problems.

6.1.6 Emissions from Manure Left on Pasture and Manure Applied to Soils

Manure (organic fertilizer) provides another alternative for farmers in SSA to improve soil fertility. The use of manure is, however, confined to play a complimentary role to synthetic fertilizers that provide significant amounts of readily available nutrients required to fuel the expansion of agriculture on the subcontinent (Richards 2016).

Current and projected emission levels from manure applied to soil and manure left on field have East Africa leading in emissions followed by West Africa. Projections (2030 estimates) of emissions from these two sources are expected to remain relatively same as current levels. The focus on integrated soil fertility management is, therefore, important to keep emission levels from synthetic fertilizer soil amendments to the minimum (Fig. 6.5).

6.1.7 Burning of Savannah and Crop Residues

Burning of savannah, a common practice in sub-Saharan Africa, involves the setting of fires to burn trees cut from forests for the development of agricultural lands. Fire is also set on existing agricultural land for nutrient mobilization, pest control and the removal of brush and litter accumulation (Ten Hoeve et al. 2012). Greenhouse gas emissions from burning of savannah consist of methane (CH_4) and nitrous oxide (N_2O) gases produced from the burning of vegetation biomass in the following five land cover types: savannah, woody savannah, open shrublands, closed shrublands and grasslands (FAOSTAT 2016) (Fig. 6.6).

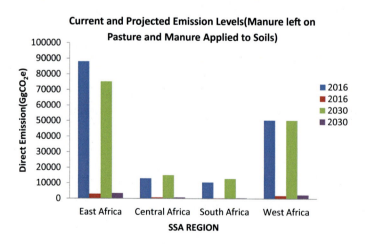

Fig. 6.5 Manure emissions from SSA (FAOSTAT 2016)

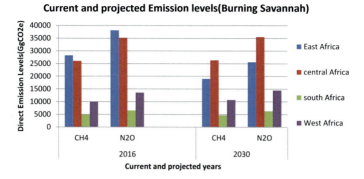

Fig. 6.6 2016 and 2030 emission levels from SSA savannah burning (FAOSTAT 2016)

Crop residues are generated in large quantities on the field seasonally after harvest. Typical crop residues include cereal straws, husks, leaves, semi-woody and woody stalks. Significant crop residue is also generated when farm produce is processed by milling. Many regions in sub-Saharan Africa use these crop residues for various purposes such as feed for animals, fuel for domestic as well as industrial use and also as thatch to roof rural homes. However, a significant amount of crop residues is left on farms whose disposal poses a great challenge for farmers in SSA. Burning of these residues on the field, therefore, presents a cheap and inexpensive way to get rid of the volumes of residues left on their farms after harvest in preparation for the new season. Greenhouse gas (GHG) emissions from the burning of crop residues consist of methane (CH_4) and nitrous oxide (N_2O) gases produced by the combustion of a percentage of crop residues burnt onsite. Air pollutants (CO_2 NH_3, NOx, SO_2, NMHC, volatile organic compounds), particulates matter and smoke are also produced as a result of the burning, thereby posing threat to human health (Jain et al. 2014).

The emissions from two important crops in SSA, maize and rice are represented below. Current emission levels (2016) indicate high methane emissions from maize

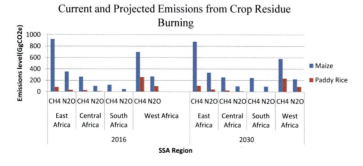

Fig. 6.7 Current and projected emission levels (2016 and 2030). *Source* (FAOSTAT 2016)

residue burning at 923 $GgCH_4CO_2e$ and 698 $GgCH_4CO_2e$ for East and West Africa, respectively. The year 2030 projections for methane emissions for the two regions, East and West Africa, are expected to slightly drop to 879 $GgCO_2e$ and 579 $GgCO_2e$, respectively (Fig. 6.7).

6.2 Adaptation and Resilience Building

The Intergovernmental Panel on Climate Change (IPCC) defines adaptation from two perspectives, human systems and natural systems. In human systems, adaptation is defined as the process of adjustment to actual or expected climate and its effects, in order to moderate harm or exploit beneficial opportunities and in natural systems, the process of adjustment to actual climate and its effects; human intervention may facilitate adjustment to expected climate (IPCC 2012). Efforts and strategies aimed at aiding SSA adapt effectively to climate risks hampered by a weak adaptive capacity of dominant smallholder farmers in the region (Parry et al. 2007). By adaptive capacity, reference is made to the strengths, attributes and resources available to an individual, community, society or organization that can be used to prepare for and undertake actions to reduce adverse impacts, moderate harm or exploit beneficial opportunities (IPCC 2012). Assisting smallholder farmers in SSA to strengthen their social, economic and ecological resilience will enable them to effectively adapt.

Resilience is the ability of a system to deal with stresses and disturbances, while retaining the same basic structure and ways of functioning, capacity for self-organization and capacity to learn and adapt to change (Field et al. 2012). In view of the above definition for resilience, (Stringer et al. 2012), for adaptation to be effective in countering the adverse effects of climate change in SSA, efforts should be directed at making adaptation strategies resilient, i.e. adaptations that can stand the test of current and future climate risks.

Sub-Saharan Africa has to imminently deal with the current climate risk it faces, while preparations are made to deal with the predicted future climate scenarios. In the light of the above, a two-way climate adaptation approach is proposed. First, it is important to have **a coping adaptation strategy** to deal with imminent risks faced by farmers. For the long term, effective **adaptation strategies** need to be developed to deal with evolving future climate scenarios that have not been experienced yet by farmers (Cooper 2013; Burton and van Aalst 2004).

By coping adaptation strategies, we refer to those strategies that farmers have employed to deal with climate stressors overtime and have understood its application and effectiveness, whereas adaptation strategies herein refer to long-term strategies that have to be developed, tested and introduced to farmers through effective extension delivery services (World Bank 2010).

Many SSA governments have developed policy documents to guide the strategic steps needed to be taken to counter climate change impacts on their economies. For example, Ghana has developed the National Climate-Smart Agriculture and Food

Security Action Plan (2016–2020) whereas Nigeria has the National Adaptation Strategy and Plan of Action on Climate Change.

A major issue that has slowed down the implementation of most adaptation initiatives in SSA has been financing. It is estimated that an annual cost of at least $18 billion is needed for adaptation to climate change programmes in SSA between 2010 and 2050 and this is exclusive to funding necessary to put SSA in the low carbon development category (Nakhoda et al. 2011). There is a general consensus that the level of financing currently reaching African countries is nowhere near enough to meet demonstrated needs, especially for immediate adaptation measures (Richards et al. 2018).

6.3 Conclusions

GHG emission levels from SSA agriculture remain low compared to other regions of the world. However, with a projected increase in intensification of agriculture to boost food production on the subcontinent, emissions especially of N_2O are most likely to surge. The surge will be attributable to the expected increase in synthetic fertilizer use for crop production. Strategies should, therefore, be centred on nitrogen use efficiency as this will ensure low emissions and also reduce production cost for already constrained smallholder farmers.

Quantification of in situ GHG emission climate change and its impact on the continent must be collaborative.

For sub-Saharan Africa to be able to cope with current and future climate risks, robust climate financing sources should be developed to sustainably fund the process of climate adaptation strategies. These financial resources when acquired should not be utilized to isolate farmers, but rather substantial investments made in their activities to improve their production efficiency and consequently reduce emissions (Cooper et al. 2008). To ensure robustness of quantified emission levels from SSA, there is the need to standardize emission quantification methodologies and wean emission research in SSA of the emission factor method, which currently dominates research on emission measurements. Efforts should, therefore, be geared towards training researchers on the use of the low-cost closed chamber, gas chromatograph method in emission measurement research on the subcontinent.

References

Alexandratos N, Bruinsma J (2012) World agriculture towards 2030/2050: the 2012 revision, vol. 12, no. 3. ESA Working paper, FAO, Rome

Awulachew SB, Merrey DJ, Kamara AB, Koppen BV, Vries FP, Boelee E (2005) Experiences and opportunities for promoting small-scale/micro irrigation and rainwater harvesting for food security in Ethiopia. International Water Management Institute (IWMI), Colombo, Sri Lanka

Bewket W, Conway D (2007) A note on the temporal and spatial variability of rainfall in the drought-prone Amhara region of Ethiopia. Int J Climatol 27:1467–1477

Blanc É (2011) The impact of climate change on crop production in Sub-Saharan Africa

Burton I, van Aalst M (2004) Look before you leap: a risk management approach for incorporating climate change adaptation into World Bank operations, World Bank Monograph, Washington (DC), DEV/GEN/37 E

Cooper PJM et al (2013) Climate change adaptation strategies in Sub-Saharan Africa: foundations for the future. InTech Open

Cooper PJM, Dimes J, Rao KPC, Shapiro B, Shiferaw B, Twomlow S (2008) Coping better with current climatic variability in the rain-fed farming systems of sub-Saharan Africa: an essential first step in adapting to future climate change? Agr Ecosyst Environ 126:24–35

FAO (2009) FAOSTAT. Food and Agriculture Organization of the United Nations (FAO) (FAO Statistical Databases)

FAO (2011) The Global Food Losses report

FAOSTAT (2016) United Nations Food and Agricultural Organization: Agricultural Data available on the World Wide Web. http://www.fao.org/faostat/en/#data

FAOSTAT Emissions database (2016). http://www.fao.org/faostat/en/#data/GH

Field CB, Barros V, Stocker TF, Dahe Q (eds) (2012) Managing the risks of extreme events and disasters to advance climate change adaptation: special report of the intergovernmental panel on climate change. Cambridge University Press

Herrero M, Thornton P, Kruska R, Reid R (2008) Systems dynamics and the spatial distribution of methane emissions from African domestic ruminants to 2030. Agr Ecosyst Environ 126:122–137

Hogarth JR, Haywood C, Whitley S (2015) Low-carbon development in sub-Saharan Africa

IMF (International Monetary Fund) (2012) International jobs report. Economist Intelligence Unit, Washington, DC

IPCC (2012) Glossary of terms. In: Field CB, Barros V, Stocker TF, Qin D, Dokken DJ, Ebi KL, Mastrandrea MD, Mach KJ, Plattner G-K, Allen SK, Tignor M, Midgley PM (eds) Managing the risks of extreme events and disasters to advance climate change adaptation. A special report of working groups I and II of the intergovernmental panel on climate change (IPCC). Cambridge University Press, Cambridge, UK, and New York, NY, USA, pp 555–564

IPCC (2014) Climate Change 2014 Synthesis Report Summary Chapter for Policymakers

Jain N, Bhatia A, Pathak H (2014) Emission of air pollutants from crop residue burning in India, pp 422–430. https://doi.org/10.4209/aaqr.2013.01.0031

Jerven M, Duncan ME (2012) Revising GDP estimates in sub-Saharan Africa: lessons from Ghana. Afr Stat J 15:13–24

Kim D-G et al (2015) Reviews and syntheses: greenhouse gas emissions in natural and agricultural lands in sub-Saharan Africa: synthesis of available data and suggestions for further studies. Biogeosci Discuss 12(19):16479–16526. https://doi.org/10.5194/bgd-12-16479-2015

Liang XQ et al (2013) Nitrogen management to reduce yield-scaled global warming potential in rice. Field Crops Res 146:66–74. https://doi.org/10.1016/j.fcr.2013.03.002

Linquist BA et al (2012) Fertilizer management practices and greenhouse gas emissions from rice systems: a quantitative review and analysis. Field Crops Res 10–21. https://doi.org/10.1016/j.fcr.2012.06.007

Lobell DB, Schlenker W, Costa-Roberts J (2011) Climate trends and global crop production since 1980. Science 333(6042):616–620. https://doi.org/10.1126/science.1204531

Ministry of Food and Agriculture (2013) Facts and figures (2012)

Morton JF (2007) The impacts of climate change on smallholder and subsistence agriculture. Proc Natl Acad Sci USA 104:19680–19685

Müller C et al (2011) Climate change risks for African agriculture. PNAS 108(11):4313–4315. https://doi.org/10.1073/pnas.1015078108

Nakhoda S, Caravani A, Bird N (2011) Climate finance policy brief climate finance in Sub-Saharan Africa

Niang I, Ruppel OC, Abdrabo MA, Essel A, Lennard C, Padgham J, Urquhart P (2014) Africa. In: Climate change 2014: impacts, adaptation and vulnerability. Contribution of working group II to the fifth assessment report of the intergovernmental panel on climate change. Cambridge University Press, Cambridge

OECD/FAO (2016) Agriculture in Sub-Saharan Africa: Prospects and challenges for the next decade. In: OECD-FAO Agricultural Outlook 2016–2025, OECD Publishing, Paris. http://dx.doi.org/10.1787/agr_outlook-2016-5-en

Parry ML, Canziani OF, Palutikof JP, van der Linden PJ, Hanson CE (2007) Climate change 2007: impacts, adaptation and vulnerability. Cambridge University Press, Cambridge, UK

Richards M et al. (2016) Fertilizers and low emission development in sub-Saharan Africa (November)

Richards MB, Wollenberg E, Van Vuuren D (2018) National contributions to climate change mitigation from agriculture: allocating a global target. Climate Policy (Taylor & Francis) 0(0): 1–15. https://doi.org/10.1080/14693062.2018.1430018

Schlenker W, Lobell DB (2010a) Robust negative impacts of climate change on African agriculture. Environ Res Lett 5(1):014010. https://doi.org/10.1088/1748-9326/5/1/014010

Schlenker W, Lobell DB (2010b) Robust negative impacts of climate change. Environ Res Lett 5(414010):8. https://doi.org/10.1088/1748-9326/5/1/014010

Schlenker W, Roberts MJ (2009) Nonlinear temperature effects indicate severe damages to U.S. crop yields under climate change. Proc Natl Acad Sci 106(37):15594–15598. https://doi.org/10.1073/pnas.0906865106

Serdeczny O et al (2016) Climate change impacts in Sub-Saharan Africa: from physical changes to their social repercussions. Reg Environ Change (Feb):1–16. https://doi.org/10.1007/s10113-015-0910-2

Signor D, Eduardo C, Cerri P (2013) Nitrous oxide emissions in agricultural soils: a review 1. Pesqui Agropecu Trop Goia 2013:322–338

Speranza CI (2010) Resilient adaptation to climate change in African agriculture. Bonn

Stringer LC, Dougill AJ, Thomas AD, Spracklen DV, Chesterman S, Speranza CI, Rueff H, Riddell M, Williams M, Beedy T, Abson DJ (2012) Challenges and opportunities in linking carbon sequestration, livelihoods and ecosystem service provision in drylands. Environ Sci Policy 19:121–135

Syakila A, Kroeze C (2011) The global nitrous oxide budget revisited. Greenh Gas Meas Manag (Taylor & Francis Group)

Ten Hoeve JE et al (2012) Recent shift from forest to savanna burning in the Amazon Basin observed by satellite. Environ Res Lett 24020(7):8. https://doi.org/10.1088/1748-9326/7/2/024020

Tollens E (2006) Markets and institutions for promoting rice as a tool for food security and poverty reduction in sub-Sahara Africa. Afr Crop Sci J 15:237–242

World Bank (2010) The economics of adaptation to climate change

World Bank (2015) Regional dashboard: poverty and equity, Sub-Saharan Africa. http://povertydata.worldbank.org/poverty/region/SSA

World Bank (2016) Agricultural Data available on the World Wide Web. https://data.worldbank.org/indicator/AG.CON.FERT.ZS?locations=ZG-CN&name_desc=false

Xia L et al (2016) Greenhouse gas emissions and reactive nitrogen releases from rice production with simultaneous incorporation of wheat straw and nitrogen fertilizer. Biogeosci Discuss (Jan):1–39. https://doi.org/10.5194/bg-2015-620

Yéo WE et al (2016) Vulnerability and adaptation to climate change in the Comoe River Basin (West Africa). Springer International Publishing, SpringerPlus. https://doi.org/10.1186/s40064-016-2491-z

You LZ (2008) Africa infrastructure irrigation investment needs in sub-Saharan Africa

Chapter 7
The UK Path and the Role of NETs to Achieve Decarbonisation

Rafael M. Eufrasio-Espinosa and S. C. Lenny Koh

Abstract The UK is one of the industrialised countries that committed to reducing its basket of greenhouse gas emissions under the Kyoto Protocol. Provisional figures in the latest inventory show a significant decrease in CO_2 emissions since 1990. This seems to indicate that the United Kingdom (UK) is on a steady transition towards a low-carbon economy. However, the ambitious goal of reducing its emissions to 80% by 2050 seems to be very difficult to accomplish. The objective of this chapter is to present a review of the current state of the greenhouse gas emissions in the UK and describe what factors are influencing the recent decline of CO_2 levels. We also provide a general overview of the strategies leading to the transition into a low-carbon economy and the sectors contributing to this process. Finally, we explore whether negative emissions technologies are ready for implementation in UK's decarbonisation path.

Keywords UK's carbon targets · Low economy · Negative emissions technologies

7.1 Introduction

In accordance with the objectives of the Paris agreement, 2050 will be a critical year to achieve global decarbonisation, and many of the international efforts to mitigate the potential impact of climate change are already in progress. The ultimate goal is to limit the increase of global warming temperatures by less than 2 °C for that year.

R. M. Eufrasio-Espinosa (✉) · S. C. Lenny Koh (✉)
Advanced Resource Efficiency Centre (AREC), The University of Sheffield,
Management School, Conduit Rd, Sheffield S10 1FL, UK
e-mail: r.eufrasio@sheffield.ac.uk

S. C. Lenny Koh
e-mail: s.c.l.koh@sheffield.ac.uk

© Springer Nature Singapore Pte Ltd. 2019
N. Shurpali et al. (eds.), *Greenhouse Gas Emissions*, Energy, Environment,
and Sustainability, https://doi.org/10.1007/978-981-13-3272-2_7

The UK is committed to achieving this goal by taking climate change actions underpinned by a broad set of environmental policies.

Provisional figures in the latest UK's inventory show 42% decrease of greenhouse emissions in comparison with the baseline year 1990, while the gross domestic product has been rising during the same period. The progress achieved so far has not gone unnoticed internationally. However, in the light of Brexit, more actions are required to meet self-imposed long-term reduction targets.

In this chapter, we present a review to explore some of the factors involved in this achievement. Although literature and reports related to UK climate change actions and policies are extensive, it is not the purpose of this revision to provide a deep understanding of them, which would require a more complete analysis. Instead, this review aims to provide a general perspective by putting together pieces of information that could be taken as potential indicators to answer if the UK is on the path to achieve decarbonisation and whether removal technologies are going to play a part of it.

For this review, we have consulted the latest 2017 annual provisional emissions results published by the Department for Business, Energy and Industrial Strategy and the most recent reports, books and scientific papers that refer to the greenhouse emission reduction progress in the UK. The components of this review are organised in the following order:

- The current state of GHG,
- Sector decarbonisation pathway,
- The transition to a low-carbon economy,
- The potential role of low-carbon and emissions removal technologies.

The final section highlights the main points of this review by contrasting strengths and weakness found in this work.

7.2 The Current State of GHG

7.2.1 UK GHG Mitigation Targets

Despite the scepticism still existing by some governments, most countries support the arguments presented by the scientific community, which has repeatedly suggested that in addition to some natural factors, climate change and global warming are the results of the significant increase of greenhouse gases (GHGs) in the atmosphere (IPCC-AR5 2015). The basket of greenhouse gases includes water vapour (H_2O), nitrous oxide (N_2O), methane (CH_4), ozone (O_3) and carbon dioxide (CO_2) found in nature, while halocarbons, hydrofluorocarbons (HFCs) and perfluorocarbons (PFCs) are usually gases products of human activities (EPA 2016).

The greenhouse effect is a widely known natural phenomenon. Basically, a thin layer of gases reflects part of the solar radiation into space, and another part is absorbed and lets the rest pass to the earth's surface. Upon being heated, the earth's surface emits longwave radiation, a part of which is trapped by the gases in the atmosphere and sent back to the earth's surface. The direct effect is the warming of this and the troposphere, for hundreds of thousands of years; this terrestrial regulatory system has maintained the average annual temperature of the planet at 15 °C allowing the evolution of life as we know it (Romm 2016). However, with the onset of the industrial revolution, the growth of the global population, the demand for more energy and the use of fossil fuels, humankind has been emitting millions of extra tons of GHG into the atmosphere, modifying this natural cycle and resulting in a significant increase in global temperature of almost 0.8 °C since 1850 (Hawkins et al. 2017).

The Kyoto Protocol is an international agreement created in the 1990s to reduce greenhouse gas emissions and limit the possible impacts that the increase of 2 °C in temperature can have on our ecosystem. In this agreement, some of the most developed countries voluntarily committed themselves to take obligatory measures to reduce the emission of these gases. The UK is one of the industrialised nations within the European Union that has committed locally, regionally and internationally to reduce its basket of greenhouse gases by considering the limits established on the Kyoto Protocol (ECC 2015).

Consistent with the long-term objectives of the first protocol, the UK has ratified its responsibility on subsequent occasions; in the second Kyoto Protocol period 2013–2020, the Climate Change Act 2008, the European Union (EU) Emissions Trading System (ETS) and more recently in the Paris Agreement 2015. Within the "load distribution" in the European Union, the UK set the initial objective of reducing its emissions by at least 12.5% between 2008 and 2012, based on the baseline 794.16 (MtCO$_2$e) levels of 1990. Domestically, The Climate Change Act 2008 initially established to reduce emissions by 60% by 2050 which was subsequently amended with a more ambitious 80% goal and with an intermediate aim by 34% by 2020 concerning the base year 1990.

With the Act 2008 scheme, the UK was the first country in the world to legally restrict the total amount of greenhouse gas emissions through carbon budgets in periods of 5 years (DECC 2014). Currently, as a member of the European Union, the UK has a jointly 20% emission reduction target with a potential increase of up to 30% by 2020. The initial target is unconditional and approved by legislation in place since 2009. However, it is uncertain if the latter could be affected by the imminent departure of the UK form the European Union in the so-called BREXIT (ECC 2015).

7.2.2 Criticism of Reduction Targets

Both the scope and the background of reduction targets in the EU and the UK have not been free of criticism. On the one hand, a group of scientist has pointed out that more ambitious targets could be easily achieved than those currently proposed by the European Commission. According to them, the proposed framework lacks aggressive plans for implementation, and the set of "modest" targets are not in line with what science requires to keep global warming below 2 °C. Furthermore, they suggested that from a scientific perspective, the mid-term collective target adopted in 2014 to cut at least 40% of emissions by 2030 will not be enough to meet the long-term reduction of 80% by 2050 (Schiermeier et al. 2014).

On the other hand, Knutti et al. (2015) stated that despite the universal acceptance of the 2 °C warming target as a safe limit to avoid dangerous climate change, this perception could be incorrect. According to the study, no scientific assessment has correctly justified or defended the 2 °C target as a safe level of warming. They agree that global temperature is the best climate target indicator, but it is unclear what level could be considered safe. So far required emission reductions based on a 2 °C warming target have been ineffective, and it is undoubtedly clear this is not a problem that science alone can address (Knutti et al. 2015).

Although in theory, The Climate Change Act and its carbon budgets have put the UK on track for the long-term decarbonisation path, and these achievements might have been exploited politically. There is historical evidence that domestic reduction targets could be achieved somewhat in an accidental manner, with past emissions reductions actions largely following on from a policy change in other sectors (Lorenzoni and Benson 2014).

Regarding estimation methods, the UK has adopted a consumption-based emissions system, based on statistics of domestic energy demand as an official government indicator. The IPCC 2006 guidelines are the basis to compile the GHG inventory, and this accounting system has been the base of numerous reports that use it to evaluate the effectiveness of mitigation actions beyond those afforded by technology. These include evaluations of resource efficiency in climate change mitigation policy, as well as services' role in the understanding of GHG emissions drivers (Barrett et al. 2013).

UK's method to account greenhouse gases has been controversial, as incorporated carbon emissions are often not considered in the set of national carbon inventories, given that emissions from fossil fuels are generally accredited only to the country where the emissions are produced (Davis et al. 2011). In the same context, the use of only national data to set emissions targets has been criticised, given the socio-economic and environmental global inequalities inherent to this approach (Baker 2018).

A recent report assessment of the inventory estimation methods states that in the UK both data and techniques used to estimate greenhouse gas emissions are updated and revised annually by incorporating the latest scientific research. According to this, the domestic GHG inventory is assembled by using a comprehensive sectoral approach (bottom-up) methodology, which provides sector-tailored inventories in accordance with the IPCC reporting format (top-down approach). Here it is also stated that the current method includes both estimations from National Statistics on production, and imports, exports, stock changes and non-energy uses of fossil fuels (Butterfield 2017).

Furthermore, with the continuous monitoring and the integration of new methodologies employed in producing the GHG inventory, accuracy has improved, and uncertainty has consistently fallen. In fact, the report claims that the current uncertainty of 3% is the third lowest when compared internationally, only behind another two less complex economies with fewer sources of emissions. Finally, the assessment states that overall, the UK has a robust and reliable inventory and monitoring system, which is carried out by The Committee on Climate Change (CCC) to set and assess carbon budget performances.

The Committee on Climate Change is an independent non-departmental public body, which incorporates both advisory and monitoring functions to inform government and achieve a credible carbon policy over a long-time frame. This organisation was introduced through the Climate Change Act (2008) and is considered as the first environmental body of its kind. The composition of the CCC seems to be well decided in that it mainly comprises individuals from a strong academic background whose experience makes them well placed to advise the government on the many issues involved in climate change (McGregor et al. 2012). With this unique background as an institution, CCC ensures independence and provides credibility in matters of climate change policies both at national and sectoral levels.

7.2.3 The Greenhouse Gas Emissions Inventory

In addition to final annual estimations, the UK inventory also provides provisional quarterly figures by source sector published by the Department for Business, Energy and Industrial Strategy (BEIS) (Butterfield 2017). Although provisional estimates are subject to a greater range of uncertainty than final figures and these are not used to evaluate whether the UK is or not on track to meet its mitigation targets, they can provide an initial glimpse of how the UK GHG inventory is performing (Brown 1990). According to the latest provisional figures released, the whole basket of GHG emissions in 2017 was at 455.9 $MtCO_2e$, which are 42.5% below the base year 1990, while CO_2 emissions were at 367 $MtCO_2e$ and this represents a 38% decrease compared to 1990 levels. To put these figures in perspective, it has been stated that "current CO_2 levels in the UK are now as low as emissions were back in 1890" (BEIS 2017; Brief 2018).

The above results have been reflected in the fulfilment of national objectives; as shown in Fig. 7.1, the first and second carbon budgets for the periods (2008–2017) have been already completed, and everything seems to indicate that the third budget will also be reached according to the set target (BEIS 2015, 2017). However, it is still uncertain whether the two remaining carbon budgets periods (2023–2032), and the final objective of reducing emissions by 80% by 2050 can be satisfactory, given the difficulties to keep a progressive annual decrease of at least 3% in domestic emissions to achieve these purposes (Change 2017).

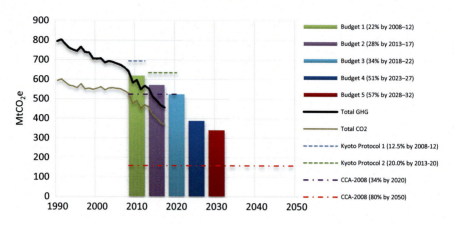

Fig. 7.1 Provisional GHG figures and UK's reduction targets (*Source* BEIS 2017)

7.3 Sectoral Decarbonisation Pathways

According to the last UK inventory, there are many factors reflected at the sectoral level involved in the reduction of GHG levels in comparison with 1990. Based on the United Nations Framework Convention on Climate Change for GHG reporting, UK's GHG inventory allocates ten high main sectors, which represent all economic activities; energy supply, business, transport, public, residential, agriculture, industrial process, land use, land-use change and forestry (LULUCF) and waste management (BEIS 2017).

As shown in Fig. 7.2 and Table 7.1, the reduction in greenhouse gas emissions in the power sector has been significantly most important than in any other

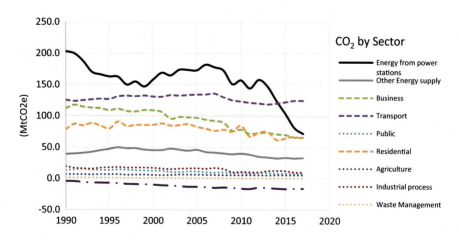

Fig. 7.2 Historical carbon dioxide emission by sector (*Source* BEIS 2017)

7 The UK Path and the Role of NETs …

Table 7.1 Current GHG and reduction percentage since 1990 (*Source* BEIS 2017)

Sector	$MtCO_2e$ (2017)	% CO_2 (2017)	% Reduction since 1990
Energy (power stations and other suppliers)	105	29	57
Business	65.8	18	41
Transport	124.4	34	1
Public	7.9	2	41
Residential	64.1	17	18
Agriculture	5.5	2	15
Industrial	9.8	3	49
Waste	0.3	0	79
LULUCF	−16.0		−4
Total CO_2	366.9	80.48	38
Rest of GHG basket	89	19.52	56
Total GHG	455.9	100	42.60

industries. After energy, transport is the sector with higher emissions with 124.4 ($MtCO_2e$), while industrial process, public, agriculture, waste management, LULUCF contribute with only 5% of the total. Residential and business are sectors with similar contributions of around 17% each, but while business has dropped its emissions to 41%, the residential sector has fallen only 18% since 1990. Sectoral factors involved in this pathway are explained below.

Overall, UK's decarbonisation path has been somewhat asymmetrical across all sectors. Now, coordinated progress across all areas will be needed to be on track to meet the intermediate fourth and fifth carbon budgets and the 2050 target. However, this concerted effort will not be easy given that the timing and level of action across sectors vary according to reduction potential, technology availability and cost-effective application among other factors (IDDRI 2015). To put this in context, a transition path, key drivers, post-Brexit scenario as well as potential challenges identified by the academic community across sectors, are described below; set out in this way some of the conditions that the UK will face in ensuring deep decarbonisation.

7.3.1 Energy Sector

Performance: Power Stations: 71.8 ($MtCO_2e$) annual CO_2 emissions during 2017 from this sector accounted for 20% of the total; there has been a 65% reduction in power station emissions since 1990. Other Energy supply: 33.2 ($MtCO_2e$) annual CO_2 emissions during 2017 from this sector accounted for 9.5% of the total, there has been a 15% reduction in other energy sources emissions since 1990. *Carbon path*: Overall, the primary cause of UK's emissions steady declines is because the energy supply sector has significantly been reduced the use of coal and gas, while has increased the use of renewables for electricity generation (BEIS 2017).

A substantial step in the systematic and complicated process of decarbonisation was the closure of Longannet and Ferrybridge in 2016, which were the largest coal suppliers in the UK, it took only 2 years to reach record levels of up to 89% in the reduction of coal use in power stations. The total closure of coal-fired power stations for electricity generation has been scheduled by 2025 (Brief 2017). *Policies contribution*: Alongside other factors UK carbon policies have contributed in this transition with a carbon price limit scheme put in place in 2013. EU policies have helped in the UK with the introduction of renewables and air quality regulations which have discouraged use of coal. *Post-Brexit actions*: The UK will try to match carbon prices with EU countries to ensure UK generators are not disadvantaged and assuring electricity interconnection (CCC 2016).

Challenges: Policies and actions in the decarbonisation of the energy sector are essential not just in Britain but all industrialised countries pursuing a low-carbon transition. Although some low-carbon options are already available across all the regions in the UK, they are not always cheap (Fankhauser 2012). In this respect, future technology choices in the power sector bring with them strong regional implications for future investment targeting, suggesting the possibility of there being regional winners and losers under different transitions. The difference between the four different UK government administrations in future energy policy decisions may create tensions about regional economic development and how an equitable energy transition can be achieved for all (Li et al. 2016).

On the other hand, it is also stated that complex decarbonisation policies have been contradicted by other measures and not efficient at all. For example, as a result of the policies aimed to reduce greenhouse gas emissions in the energy sector, the increasing cost of electricity up to 50% in UK households during the last 17 years is part of the side effects. This is because electricity prices oscillate according to fuels' prices. Finally, it is suggested that if some inclusive environmental–social policies would have been considered, emissions reduction targets could be achieved at a much lower cost (Zuluaga 2017).

7.3.2 Residential and Public Sectors

Residential Performance: 64.1 ($MtCO_2e$) annual CO_2 emissions during 2017 from this sector accounted for 17% of the total; there has been an 18% reduction in the residential sector emissions since 1990. Carbon emissions have dropped in this sector; however, warmer climate conditions during this period might have contributed to reducing the use of natural gas for heating, which along with cooking are the primary sources of emissions in this sector (BEIS 2017).

Public Performance: 7.9 ($MtCO_2e$) annual CO_2 emissions during 2017 from this sector accounted for 2% of the total; there has been a 41% reduction in the public sector emissions since 1990. *Carbon path*: The use of natural gas for heating public buildings is the central source of emissions from this sector. During the last 15 years, there has been a dropping emissions tendency in the public sector motivated by a change in the fuel mix (with more use of natural gas and less use of coal and oil) (BEIS 2015).

Policies contribution: In spite of that the population has grown and incomes have risen, UK-EU energy efficiency standards in new buildings and regulations on new boilers have helped cut residential emissions by reducing demand for energy. *Post-Brexit*: Currently, there are some policies agreed between the UK and EU to reduce emissions from buildings. The target of these policies is the use of renewables, products labelling and improving energy efficiency. After Brexit, UK will be focused on low-carbon heat, will replicate EU energy efficiency and labelling standards at UK level, and will improve buildings insulation (CCC 2016).

Challenges: Public Buildings. A 22% (19 $MtCO_2e$) reduction by insulating 7 million lofts, cavity, and solid walls and increasing take-up of low-carbon heating (i.e. heat pumps or district heating) to about 4 million homes and around half of the non-residential buildings. In both the residential and the public sectors are anticipated that new low-carbon heat technology and district heating will play an essential role for decarbonisation. Gas-based central systems dominate heating in UK's residential industry, and this will remain for at least ten more years. A report published by the Institute for Sustainable Development and International Relations points out that increasing penetration of hybrid heat pumps/gas systems will be observed by 2030. However, a radical shift away from gas use will be challenging, with potential economic implications for gas distribution systems and how they might be operated (IDDRI 2015).

In the same context, a journal article has highlighted that in spite of that the use of natural gas will reduce over time, this will play only a secondary role in helping meet targets based on heat demand. According to this article, heating is perhaps one of the most challenging aspects to decarbonise in UK's energy system given that high levels of uncertainty about how heat will be decarbonised present some challenges to policymakers (Chaudry et al. 2015).

7.3.3 Transport

Performance: 124.4 ($MtCO_2e$) annual CO_2 emissions during 2017 from this sector accounted for 34% of the total, there have been only a 1% reduction in the transport sector emissions since 1990. *Carbon path*: in this sector emissions have been broadly flat, and road transport has been the most critical cause of emissions despite the increase in diesel consumption and lower petrol consumption, (BEIS 2017). *Policies contribution*: UK policies such as "Carbon price floor" have contributed significantly to reduce coal use in electricity generation, while EU policies have improved the efficiency of new cars since 2009; however, during this period the increase of distances travelled has offset emissions reductions. *Post-Brexit*: The propose is to remain as a part of the EU vehicle efficiency standards or replicate these at the UK level, preserve EU regulations that encourage efficient driving, and maintain or replicate the carbon impact of broader EU-led legislation to incentivise uptake of cleaner vehicles such as the Air Quality Framework and National Emission Ceilings Directive (CCC 2016).

Challenges: The UK will target transport's emissions reductions by 43% below 2015 levels by 2030.

In the light of Brexit, innovation policies in the transport sector should be strongly considered. A study by Byers et al. (2015) suggests improving policy-relevant insights about the sustainability of multiple future transport pathways by introducing energy analysis as a measurement. In this study, the current set of transport policy options to reduce national GHG emissions was used. According to the authors, the principal aim of these policies is based on transport efficiency, and it is conventionally measured as the product of goods/passengers, but it does not reflect potential sectoral improvements both regarding technical efficiency and resource use efficiency of the energy sector.

In addition, Upham et al. (2013) it is also suggested that current expectations and visions of transport system innovation are still very much focused on motor vehicle technology change. However, there are no signs of acceptance of transport demand reduction policies, which in turn may be regarded as a form of social innovation. Finally, a newspaper article (Gabbatiss 2018) states that within all the UK sectors, is transport which is "failing to play its part" in restricting emissions, making it the worst performing sector. Here is also reported that a high ranking member from the European Parliament's Transport and environment committees has criticised the "wrong-headed" UK priorities of cutting bus services, de-prioritising walking and cycling infrastructure while discarding plans for rail electrification.

7.3.4 Business and Industrial Process

Business: *Performance*: 65.8 (MtCO$_2$e) annual CO$_2$ emissions during 2017 from this sector accounted for 18% of the total; there has been a 41% reduction in the business sector emissions since 1990. *Carbon path*: The business sector includes emissions from combustion in industrial/commercial sectors, industrial off-road machinery, and refrigeration and air conditioning, in this way most of the Fluorinated gases (F-gases) emissions come from this sector. Emissions in this sector have dropped by economic factors and globalisation. Overall, the primary driver has been a reduction in industrial combustion from iron and steel; however, emissions from F gases have increased considerably (BEIS 2017).

Industrial Process: *Performance*: 9.8 (MtCO$_2$e) annual CO$_2$ emissions during 2017 from this sector accounted for 3% of the total; there has been a 49% reduction in the industrial sector emissions since 1990. *Carbon path*: emissions from this sector have decreased mainly driven by globalisation of industrial production. The lower manufacturing output from cement production in the last decade has been a fundamental factor of this reduction (BEIS 2017). *Policies contribution*: EU policies EU Emissions Trading System, Renewable Energy Directive, and Energy Efficiency Directive are current measurements helping the industry reduction targets. These policies allow the UK-based industry to purchase emissions reduction from abroad where reducing emissions in the UK would be more expensive. *Post-Brexit*: UK policy should aim to reduce industry emissions by 23% below 2015 levels by 2030; however, leaving the EU might affect how this is delivered:

For example; If the UK leaves the EU, then new strategies will be needed to secure industry incentives in energy efficiency and to develop new low-carbon technologies. Alongside these measurements, the country should also continue with its current low-carbon heat and energy efficiency strategies in this sector (CCC 2016).

Challenges: A Life Cycle Assessment study (LCA) carried out by Barrett et al. (2018) point out that in spite that the cement production has been a factor in emissions reduction, now that there are no wet kilns left in the UK, further potential for reducing energy demand in this way is limited. Aligning with previous studies about the potential impact of embodied emissions in the UK, and these authors agreed that they should also be carefully considered, given that the required energy and associated GHG emissions at different points along these UK supply chains emanate from many different countries, due to the growth of globalisation. In the same context, it is identified that UK's business and industrial sectors are complex and carbon accounting is not an easy task. Here it is also recognised that data availability on industrial energy use and its potential for GHG emissions reduction is arguably the weakest in comparison to the rest UK end-use demand sectors. It is requested better information in support of the industrial modelling needs of UK policymakers (Griffin 2016).

7.3.5 Agriculture

Performance: 5.5 (MtCO$_2$e) annual CO$_2$ emissions during 2017 from this sector accounted for 2% of the total; there has been a 15% reduction in the agriculture sector emissions since 1990 (BEIS 2017). *Carbon path*: The decrease in emissions in agriculture has been driven by a decline in synthetic fertiliser utilisation and a fall in animal numbers over the period. Unlike other sectors, methane and nitrous oxide related to livestock and the use of fertilisers on agricultural soils are main sources of emissions in this sector. In this sector, carbon dioxide emissions are related to stationary combustion sources and off-road machinery (BEIS 2015). *Policies contribution*: Although there are no direct EU policies, the EU Common Agricultural framework, and other EU environmental policies have helped the reduction of agriculture's emissions. *Post-Brexit*: the aim is to reduce 15% of emissions from this sector below 2014 levels by 2030 (CCC 2016).

A recent letter from CCC addressed to Michel Gove, MP, points out that despite the vital role of agriculture to cut carbon emissions, in comparison with other sectors, during the last 6 years there has been no progress in the reduction of GHG emissions of this area. Likewise, there has been no tangible progress in the reduction emissions intensity associated with growing crops. The potential consequences might be an increasing climate risk on the farmed countryside and upland peatlands. In this letter has also been highlighted the need for a robust local and national data to be able to evaluate the impact of actions and to monitor improvements over time (CCC 2017a).

7.3.6 Waste Management

Performance: 0.3 ($MtCO_2e$) annual CO_2 emissions during 2017 from this sector accounted for 0.08% of the total; there has been a 79% reduction in the waste management sector emissions since 1990. *Carbon path*: overall emissions in this sector have dropped mainly due to reductions in the amount of biodegradable waste landfill, reducing in this way methane emissions arising from landfill sites (BEIS 2017). *Policies contribution*: the UK landfill tax has been the primary driver for this sector emissions reductions. This tax was introduced to meet obligations agreed between the UK and EU through EU's landfill directive. *Post-Brexit*: UK zero or low levels of waste to landfill will be the primary target by 2025 by keeping the landfill tax and by launching incentives/requirements for the separate collection of biodegradable waste streams. *Challenges*: The UK should aim to reduce waste emissions by 44% below 2014 levels by 2030. Leaving the EU may affect how this will be delivered. However, in spite of that additional targets have been proposed, so far they have not yet agreed (CCC 2016).

The CCC has identified potential opportunities related to methane capture, flaring and energy recovering in the waste sector. However, given that the exact number of landfill sites in the UK is still unknown, it has been complicated to deliver strategies for emissions reduction. The uncertainty in the number of landfill sites might also compromise the accuracy in the current accounting of GHG emission in the waste sector (CCC 2017b).

7.3.7 LULUCF (Land Use, Land-Use Change, and Forestry)

Performance: −16.0 ($MtCO_2e$) annual CO_2 emissions during 2017 (based on 2016 statistics) from this sector accounted for −4% of the total; there has been a 79% reduction in the LULUCF sector emissions since 1990. *Carbon path*: Overall, land-use change to cropland is the most significant source of CO_2 emissions, and remaining forest land is the dominant source of the sink. Although there has been a reduction in emissions due to less intensive agricultural practices, overall emissions have slightly increased. *Post-Brexit*: The UK approach to reducing emissions from this sector depends on regulation negotiated and agreed with EU and is expected that this will continue to cover all of the reduction required by 2030. In addition, it is also anticipated that this sector will remain as a net carbon sink beyond 2050. *Challenges*: the aim in this sector will increase afforestation to 15,000 hectares per year, reduce the horticultural use of peat and peatland restoration.

Similarly, for waste management, LULUCF needs to provide accurate estimates given that there is a high degree of uncertainty associated with the carbon estimates obtained. Therefore, in both cases spatially specific analytical approaches could be of value to policymakers to identify areas where land-use changes have higher carbon emission and sequestration rates. The spatial identification in this sector will make a significant contribution to the UK in meeting part of their emissions reduction targets and would contribute most to increasing carbon storage by conserving existing carbon stocks (Cantarello et al. 2011).

7.4 The Transition to a Low-Carbon Economy (LCE)

Taking into consideration official estimates of 1990–2017 UK, the intentional community has highlighted greenhouse gas reduction achievements; Last year a Spain's newspaper published an article entitled "The British recipe for disengaging coal" when is stated that among the great powers of Europe, the UK is a "prominent student" in cutting greenhouse gas emissions. Here, it is also pointed out that the country is making a rapid transition in coal reduction because of the CO_2 emissions tax policy imposed on power plants (Planelles 2018). A report in 2014 situated the UK among the group of countries leaders which are taking action against climate change. The report stated that in spite of that the number of sections in the UK legislation related to climate change was higher in comparison with other developed countries, tracking so many laws can be very complicated (Bassi et al. 2014).

The concept of a low-carbon economy supports the premise of "a renovated state of mind, of some governments by acting and operating to reduce their carbon intensities, through the sustainable use of resources, economic development and improvement in the quality of life" (Baranova et al. 2017). In this way, a low-carbon economy encourages activities which produce goods or services and returns low-carbon outputs (BIS 2015). Under this context, a low-carbon economy should be underpinned by a solid legal basis, putting a price on carbon, clean electricity, changes in lifestyle and behaviour are needed and is economically and technologically feasible (Fankhauser 2012).

According to the World Resource Institute (WRI 2018), the UK is an example of this concept, where during the last decade a constant economic development and greenhouse have been progressively separated. The same view is expressed in the low-carbon economic index 2017, which calculates global carbon intensities in (tCO_2/\$m GDP), and the rate of change needed in those intensities in future to limit warming to two degrees by 2100. In this report is stated that "The UK is leading the low-carbon revolution" (PWC 2017).

The low-carbon economy in the UK is defined by 17 sectors which have been assembled into six groups, which are: low-carbon electricity, low-carbon heat, energy from waste and biomass, energy-efficient products, low-carbon services and low emission vehicles. Since 2013 more than 14,000 businesses have been directly involved in the UK low-carbon economy, with more than 208,000 people working in these activities (ONS 2016).

The results of these achievements are reflected in the economic growth (GVC) within the low-carbon economy group and the entire UK economy (GDP). As shown in Fig. 7.3, during the last decade, the GDP grew from £m 1,749,216 to £m 1,959,707, while in the same period GHG emissions dropped from 640.26 to 455.89 million tonnes of carbon dioxide equivalent (MtCO$_2$e) (BEIS 2017; ONS 2018).

However, is not all rosy, a recent report (Matikainen and Druce 2018) identifies a mix of factors that are dropping the green investment in the UK. Among those factors, a reversed a number of low-carbon policies such as the elimination of renewable obligations by the UK during 2015 have "harmed confidence in new low-carbon projects". With Brexit on the horizon, the privatisation of the Green Investment Bank, which is currently one of the mains sources of UK' funds in renewables and green projects might face a significant reduction too. Finally, the authors suggested that all these elements have provoked "a dramatic collapse" in green investments during the last 2 years, which it is at its lowest level in 10 years. This is concerning, given that to meet the 4th and 5th carbon budgets by 2032 will be necessary a significant amount of financing equivalent to up to 1% of the annual GDP. Here it also stated that whether this "collapse" is a temporary dip or part of a longer-term trend is unclear.

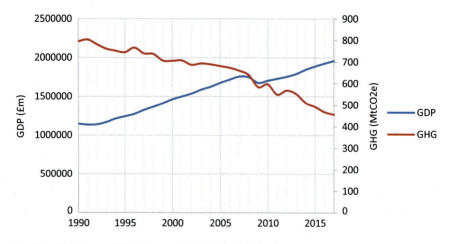

Fig. 7.3 GDP versus GHG (*Source* BEIS 2017; ONS 2018)

7.4.1 GHG Post-Brexit

Since the terms to reduce GHG emissions in the UK were defined, there was already a political inclination to comply with these obligations. However, a study carried out in 2010 already pointed out that the UK 80% reduction target "was not a realistic option if this was implemented in solitary given that it would not lead to a substantial decrease in climate change and because no single country would easily take a decision moving towards such policies on its own" (Dagoumas and Barker 2010). Now, a few years later, the UK is leaving the EU, and despite uncertainties during this process, it will take that challenge on its own.

The European Union's emissions trading system (ETS), currently regulates nearly half of UK's emissions reduction policies. A recent report (Martin 2017) points out that, within UK's political environment, all parties are aware of their commitment to continue taking action against climate change and all of them have expressed their support for the legally binding climate targets that the UK has set itself. However, according to author's opinion, this is "cheap talk" given that so far the targets have been reached relatively easy, but this was primarily due to a weak economy. Now, there is not a clear strategy in how parties are going to manage climate policies after Brexit to guarantee that the objectives are met; therefore, it is not clear that the targets will be straightforwardly met in future.

Current UK's reduction targets "carbon budgets" are based on the domestic legislation Climate Change Act 2008, which is more ambitious than those found in the EU. According to the Committee on Climate Change, after considering potential scenarios post-Brexit, "the fifth carbon budget will remain applicable whether or not the UK is a member of the EU". Therefore, UK targets for GHG emissions reduction have not changed, new policies to reinforce existent commitments will be required, and some EU policies should be preserved and improved. As the UK leaves the EU, potential impacts have been identified, and priorities and actions have been established at the sectoral level (CCC 2016).

Although during the campaign for the Brexit referendum climate change was never the main priority, now Brexit presents both challenges and opportunities to reform the climate change policy background in the UK. According to Hepburn and Teytelboym (2017) at first glance, the impact of Brexit on domestic climate policy might look to be somewhat limited. However, the potential economic and political consequences of this transition are not clear. They suggest that if the policymakers forecast are correct, it is probable that we will see a reduction of GHG emissions in short to medium term as a consequence of reduced economic activity. However, that would not necessarily be good news for climate change in the long term. Hepburn states that based on international experiences is well known that stronger, more confident, more innovative and more dynamic economies are in much better positions especially with the voters to cut emissions in the long term.

7.5 The Potential Role of Low-Carbon and Emissions Removals Technologies

We have already mentioned that until now the dramatic decrease of coal use in the electricity mix has been the main driver in the decarbonisation of the UK. While contrary to what was estimated six decades ago, nuclear energy has not played a relevant role during this process, and the potential use in the short term of greenhouse gas emissions removal technologies is still no clear.

7.5.1 Nuclear

It is well known that nuclear power is a significant means of reducing greenhouse gas emissions and in order to overcome previous results more invest in this technology is expected. Currently around 21% of the UK electricity comes from Nuclear energy, but this capacity will be reduced almost to half by 2025 given that old plants are taken offline. However, as part of the Climate Change Act, it is expected that nuclear power will play an essential role in UK's decarbonisation strategy to cut CO_2 emissions by 2050. According to the Department for Business, Energy and Industrial Strategy (BEIS), the government has plans to further develop nuclear capability by investing in more capacity and regulation frameworks (News 2017).

However, this strategy might be compromised given that there is still a general concern about cost and safety in the use of this technology. While nuclear energy is still expensive due to high funds for infrastructure projects, the cost of renewable energy has fallen significantly, making it harder for nuclear to be considered attractive. Moreover, after the Fukushima accident a few years ago, there has been a great alarm over the use of nuclear energy, and there is an increasing concern about how safe is the long-term storage of nuclear waste. These factors have caused that much attention is now being placed on greenhouses emissions removal technologies (LSE 2018).

7.5.2 Negative Emissions Technologies (NETs)

In order to maintain temperature levels below 2 °C, it is necessary to implement large-scale negative carbon technologies, in this way we will have at least one opportunity of >50% meet this goal (Smith et al. 2015). Negative emissions technologies (NETs) involve a diverse group of methods that share the same aim to withdraw GHGs from the environment, and they fit well within the UK strategy to meet long-term reduction targets whilst maintaining a competitive economy. In fact, the UK claims that already has placed indicators in some of these technologies to track progress towards meeting its domestic carbon budgets (Berg et al. 2017).

However, according to McLaren, it is not expected that all NETs will offer in the short term an economically and technical alternative for mitigation purposes. This author contemplates that perhaps these technologies would contribute to UK's decarbonisation process by 2030–2050. In order to reduce the uncertainty and to reach the potential deployment of these technologies, the appropriate regulatory framework should start now (McLaren 2012). In addition, Haszeldine et al., agree with the potential timeline deployment of NETs and also highlight the extreme cost and regulatory innovation required for climate change mitigation. They state that currently there are only a few examples of NETs applications, such as bioenergy with CCS (BECCS), and all of them have been applied only at small-scale and does not yet exist at industrial scale (Haszeldine et al. 2018).

Carbon Capture and Storage (CCS) already exists in many forms and at low cost. The significance of CCS is well known given that this technology can perform different functions, being its most significant competency sequestering carbon from the atmosphere when is used along with bioenergy. In contrast with other technologies, the flexibility and versatility of CCS offer the potential to contribute to decarbonisation through a range of diverse processes in different sectors, being the energy sector key for its application (Krey et al. 2014).

7.5.3 Carbon Capture and Storage

Carbon capture and storage (CCS) technologies process, capture and permanently store underground carbon emissions from strategic point sources, such as fossil fuel power plants, in this manner CCS allows the continued use of fossil fuels in power generation and some industrial processes. This method avoids climate-damaging effects of CO_2 emissions that otherwise would be released to the atmosphere. It is anticipated that in the UK around 90% of the CO_2 emitted from these strategic point sources could be captured and safely stored by using CCS (DECC 2012).

Within the set of IPCC scenarios, it is estimated that a total of 810 bn tonnes of CO_2 will have to be captured by 2100, however, in order to have hope, it is also suggested that the use of the CCS technologies should begin no later than 2020. Under this potential scenario, the Paris Agreement strongly suggests that countries should invest in new ways to cut emissions. So far politicians have focused efforts on curbing current emissions in Earth's atmosphere, but they have ignored the use of CCS (2017).

According to the International Energy Agency, CCS technologies will take a significant role in worldwide efforts to limit global warming, delivering at least 20% of the emissions reductions needed by 2050. To keep on track with this target and to resolve challenge more than 3,000 projects must be in progress and running by 2050. CCS is the only method that can turn high carbon fuels into genuinely low-carbon electricity, and this will be essential if we are to meet the challenge of climate change while maintaining the security of energy supplies (IEA 2013).

In the UK and its continental platform, the technical CO_2 storage capacity could be up to 70 billion tonnes, which would be enough to store 100 years' worth of current emissions from the energy sector. In this context, the Department of Energy and Climate Change (DECC 2012) established a roadmap to support the deployment of Carbon Capture and Storage technologies across the territory. The initial objectives were addressed to develop a regulatory environment to remove barriers to deployment, support the private sector by working closely with industry, and to launch a CCS commercialisation programme with £1bn of capital support. In that program, the Government claimed to make substantial investments in research, development and innovation, with some results are already showing dividends. According to the roadmap, UK's academic research is among the best in the world, international companies are willing to collaborate with UK institutions, and innovative new companies are emerging with technologies which could further diminish the cost and risk of CCS.

In theory, intellectual and available technology resources could lead to saving energy and reducing carbon dioxide emissions in the short and medium term. In fact, research and much of the technology in CCS has been available in the UK for some time. During 2017 has been proclaimed by the UK as the first country in the world to assign specific funds for local CCS technology research. However, so far it has not been the case; a study by Griffin et al. (2017) showed that the possible technological advance and its commercial use for the middle of this century are still more than speculative. According to Griffin, there are many types of obstacles that could significantly limit the potential development of CCS, and this uncertainty could have the consequence of reducing the possibilities of the UK to achieve a low-carbon economy by 2050.

In addition, it has already been pointed out (Moe and Røttereng 2018) that funds allocated for this purpose to date have been insignificant and have not received adequate attention. An example of this situation occurred in 2007 when the government launched a competition to find the low-cost bidder to build two commercial-scale plants for CCS. However, the Government cancelled the competition a few years later because it could not find an independent contractor willing to work for the money being offered (Quartz 2018). After that, the UK launched a second competition in 2012; the winners would get £1 billion. However, again in 2015, when the final bids were submitted, the government "abruptly scrapped this public grant for industrial-scale CCS plants, a grant which would have improved the implementation of this technology" (Economist 2017).

Recognising that CCS technology is at an early stage of development and has not been yet technically proven at full scale, the importance of risk factors for implementation should be urgently evaluated (Fais et al. 2016). Moe and Rottereng believe that there are apparently numerous reasons why NETs and CCS have yet not been entirely attractive, but cost and uncertainty are among the most obvious. According to them, there is a general perception of immaturity in these technologies, which makes them difficult to make plans for long-term strategies. However, for the enabling and deployment at a large scale of such controversial technologies, regulatory frameworks are needed rather than discourage them. Finally, it is stated, that currently the UK does not have plans for large-scale negative emissions, and the post-carbon society is not currently within sight (Moe and Røttereng 2018). However, if the cost or inability of selecting technology pathway such as nuclear and NETs for deployment is still uncertain, other alternative technologies such as offshore wind could reach the same level of decarbonisation a lower cost (Roberts et al. 2018).

7.6 Conclusions

As stated at the beginning of this chapter, the aim of this review attempted to provide a general perspective by putting together pieces of information that could be taken as potential indicators to answer if the UK is on the right path to achieve decarbonisation and whether emissions removal technologies are going to play a part in it. This final section highlights the main points of this review by contrasting the weaknesses and strengths found during this work.

On the one hand, both the scope and the background of reduction targets are still subject of criticism, the EU proposed framework lacks aggressive plans for implementation, and the set of "modest" targets are not in line with what science requires to keep global warming below 2 °C. Embodied carbon emissions and carbon leaks are often not considered in the set of national carbon inventories, given that emissions from fossil fuels are generally accredited only to the country where the emissions are produced. There is excessive attention by domestic energy policies to account for GHG emissions only at national scale. There is uncertainty in greenhouse emissions estimations given that there is not accurate data in some sectors, as an example of this, the exact number of landfill sites in the UK is still unknown, and this complicates the delivering of strategies.

Overall, the UK decarbonisation progress has been somewhat asymmetrical across all sectors. Now, coordinated progress across all sectors will be needed to be on track to meet the intermediate fourth and fifth carbon budgets and the 2050 target. In this context, there is not a clear strategy in how parties are going to manage climate policies after Brexit to guarantee that the objectives are met.

Brexit uncertainty is part of the factors that are dropping the green investment in the UK. Among those factors, a reversed some low-carbon policies such as the elimination of renewable obligations by the UK during 2015 have harmed confidence in current and new low-carbon projects. In this way, nuclear energy is still expensive due to high funds for infrastructure projects, and the cost of renewable energy has fallen significantly, making it harder for nuclear to be considered attractive.

Finally, despite the importance of NETs and Carbon Capture and Storage technologies (CCS) to limit global warming, they have yet not received enough attention from policymakers. Likewise, the potential contribution to a circular economy from other NETs technologies such as Carbon Capture and utilisation (CCU) has not yet considered. Overall, there is a general perception that NETs technologies are still at an early stage and have not been yet proven at full scale.

On the other hand, since the potential impact of climate change was identified as a result of the increase in greenhouse gases, the United Kingdom (UK) committed itself both internationally and domestically to establish a set of reduction targets. European shared, and UK's domestic targets are ambitious and go beyond the average measurements so far adopted by the international community. The Climate Change Act and The EU Emissions Trading System (EU ETS) are part of the environmental set of adopted policies and have been the foundations to combat climate change in the UK. Together they represent a robust legislative package that has been fundamental to comply with this goal and to meet reduction targets.

As an independent and non-departmental public body, The Committee on Climate Change (CCC) provides reliable and systematic monitoring progress of indicators formed under the Climate Change Act. This allows the UK to make plans and to take reduction actions based on the early identification of issues in any areas where targets could be missed. Current levels of total greenhouse emissions in the UK are similar to those of more than 120 years ago. The progressive reduction of 42% concerning the base year 1990 proves that as a whole the UK has consistently demonstrated an effective a solid performance. Until now there has been a steady path to decarbonisation where scheduled carbon budgets have been met. This has been reflected across the main economic sectors of the country, where the main driver has been the decarbonisation of the energy sector, while transport is the one that presents the most challenges ahead.

The transition to a low-carbon economy is on its way. There has been a growing separation between the gross domestic product (GDP) and greenhouse gases emissions (GHGs). There is a favourable recognition by the international community about the efforts and actions taken in the UK to cut down its greenhouse gas emissions. In this aspect, the UK is considered as leader country and the package of environmental policies as a model to follow. In the UK, there is a general atmosphere of uncertainty regarding the eventual exit of the European Union (Brexit). Although there are signs of distrust in the investment in green, the situation is being seen as an opportunity to reinforce the current environmental policies.

Whether the UK is or not on track to meet its decarbonisation could be subject to controversy, and indeed, it will need more in-depth analysis. In this review, the balance between the negative and the positive indicators considered above seem to suggest that in theory, the UK is on the right path to achieve decarbonisation. However, the UK has a long way to meet the 80% reduction target by 2050. In agreement with some of the references consulted in this work, this target could be highly compromised if trust, funds and regulatory frameworks are not addressed to the group of NETs in their respective timing and scales of deployment.

References

Baker L (2018) Of embodied emissions and inequality: rethinking energy consumption. Energy Res Soc Sci 36:52–60

Baranova P et al (2017) The low carbon economy: understanding and supporting a sustainable transition. In: (ed) M Palgrave

Barrett J et al (2013) Consumption-based GHG emission accounting: a UK case study. Clim Policy 13(4):451–470

Barrett J et al (2018) Industrial energy, materials and products: UK decarbonisation challenges and opportunities. Appl Therm Eng 136:643–656

Bassi S et al (2014) Walking alone? How the UK's carbon targets compare with its competitors' Centre for Climate Change Economics and Policy Grantham Research Institute on Climate Change and the Environment

BEIS (2017) 2015 UK Greenhouse Gas Emissions. Final Figures, Statistical Release: National Statistics. Department for Business, Energy and Industrial Strategy

BEIS (2018) 2017 UK greenhouse gas emissions. Provisional figures, statistical release: national statistics. Department for Business, Energy and Industrial Strategy

Berg T, Mir G-U-R, Kühner A-K (2017) CCC indicators to track progress in developing greenhouse removals options. ECOFYS

BIS (2015) The size and performance of the UK low carbon economy. The UK Department for Business, Innovation and Skills

Brief C (2018) Analysis: UK carbon emissions in 2017 fell to levels last seen in 1890. UK Emissions 2018. https://www.carbonbrief.org/analysis-uk-carbon-emissions-in-2017-fell-to-levels-last-seen-in-1890. Accessed 4 Jun 2018

Brown P et al (2018) UK greenhouse gas Inventory, 1990 to 2016. Annual Report for Submission under the Framework Convention on Climate Change

Butterfield D (2017) Understanding the UK Greenhouse Gas Inventory, An assessment of how the UK inventory is calculated and the implications of uncertainty. National Physical Laboratory. Commissioned by the Committee on Climate Change

Byers EA, Gasparatos A, Serrenho AC (2015) A framework for the energy analysis of future transport pathways: Application for the United Kingdom transport system 2010–2050. Energy 88:849–862

Cantarello E, Newton AC, Hill RA (2011) Potential effects of future land-use change on regional carbon stocks in the UK. Environ Sci Policy 14(1):40–52

CCC (2016) Meeting carbon budgets—implications of Brexit for UK climate policy. Committee on Climate Change

CCC (2017a) Letter to Michael Gove, MP. Committee on Climate Change

CCC (2017b) Meeting carbon budgets: closing the policy gap. 2017 Report to Parliament. Committee on Climate Change

Change, C.o.C (2017) Committee on climate change. How the UK is progressing. https://www.theccc.org.uk/tackling-climate-change/reducing-carbon-emissions/how-the-uk-is-progressing/. Accessed 07 Jun 2018

Chaudry M et al (2015) Uncertainties in decarbonising heat in the UK. Energy Policy 87:623–640

Dagoumas AS, Barker TS (2010) Pathways to a low-carbon economy for the UK with the macro-econometric E3MG model. Energy Policy 38(6):3067–3077

Davis SJ, Peters GP, Caldeira K (2011) The supply chain of CO_2 emissions. PNAS

DECC (2012) CCS roadmap, supporting deployment of carbon capture and storage in the UK

DECC (2014) Final Statement for the first carbon budget period. Presented to Parliament pursuant to section 18 of the Climate Change Act 2008. 2014, Department of Energy & Climate Change

ECC (2015) UK progress towards GHG emissions reduction targets, statistical release: official statistics. Department of Energy and Climate Change

Economist T (2017) Sucking up carbon, greenhouse gases must be scrubbed from the air. https://www.economist.com/briefing/2017/11/16/greenhouse-gases-must-be-scrubbed-from-the-air. Accessed 12 Jun 2018

EPA (2016) Greenhouse gas emissions, overview of greenhouse gases, United States environmental protection agency. https://www.epa.gov/ghgemissions/overview-greenhouse-gases. Accessed 07 Jun 18

Fais B et al (2016) Impact of technology uncertainty on future low-carbon pathways in the UK. Energy Strategy Rev 13–14:154–168

Fankhauser S (2012) A practitioner's guide to a low-carbon economy: lessons from the UK, Centre for Climate Change Economics and Policy Grantham Research Institute on Climate Change and the Environment

Gabbatiss J (2018) Transport becomes most polluting UK sector as greenhouse gas emissions drop overall. Independent. Accessed 16 Dec 2018

Griffin PW, Hammond GP, Norman JB (2016) Industrial energy use and carbon emissions reduction: a UK perspective. WIREs Energy Environ

Griffin PW, Hammond GP, Norman JB (2017) Industrial energy use and carbon emissions reduction in the chemicals sector: a UK perspective. Appl Energy

Haszeldine RS, Flude S, Johnson G, Scott V (2018) Negative emissions technologies and carbon capture and storage to achieve the Paris agreement commitments. Phil Trans R Soc A 376 (2119):20160447

Hawkins E et al (2017) Estimating changes in global temperature since the preindustrial period. Bull Am Meteor Soc 98(9):1841–1856

Hepburn C, Teytelboym A (2017) Climate change policy after Brexit. Oxford Rev Econom Policy 33(S1):S144–S154

IDDRI (2015) Pathways to deep decarbonization in the United Kingdom

IEA (2013) Technology roadmap, carbon capture and storage. Int Energy Agency

IPCC-AR5 (2015) Fifth assessment report, climate change 2014, synthesis report. Intergovernmental Panel on Climate Change

Knutti R et al (2015) A scientific critique of the two-degree climate change target. Nat Geosci 9:13

Krey V et al (2014) Getting from here to there—energy technology transformation pathways in the EMF27 scenarios. Clim Change 123(3):369–382

Li FGN, Pye S, Strachan N (2016) Regional winners and losers in future UK energy system transitions. Energy Strat Rev 13–14:11–31

Lorenzoni I, Benson D (2014) Radical institutional change in environmental governance: Explaining the origins of the UK Climate Change Act 2008 through discursive and streams perspectives. Glob Environ Change 29:10–21

LSE (2018) What is the role of nuclear power in the energy mix and in reducing greenhouse gas emissions? The Grantham Research Institute on Climate Change and the Environment. http://www.lse.ac.uk/GranthamInstitute/faqs/role-nuclear-power-energy-mix-reducing-greenhouse-gas-emissions/. Accessed 18 Jun 2018

Martin R (2017) Brexit as climate policy: the agenda on energy and the environment. Centre for Economic and Performance, L.S.o.E.a.P.S. (LSE), Editor

Matikainen S, Druce V (2018) Why is low-carbon investment dropping in the UK? Grantham Research Institute on Climate Change and the Environment. http://www.lse.ac.uk/GranthamInstitute/news/why-is-low-carbon-investment-dropping-in-the-uk/. Accessed 16 Dec 2018

McGregor PG, Swales J, Winning MA (2012) A review of the role and remit of the committee on climate change. Energy Policy 41:466–473

McLaren D (2012) A comparative global assessment of potential negative emissions technologies. Process Saf Environ Prot 90(6):489–500

Moe E, Røttereng J-KS (2018) The post-carbon society: rethinking the international governance of negative emissions. Energy Res Soc Sci 44:199–208

News WN (2017) UK's clean growth strategy highlights nuclear. http://www.world-nuclear-news.org/NP-UKs-Clean-Growth-Strategy-highlights-nuclear-12101702.html. Accessed 27 Jun 2018

ONS, UK environmental accounts: low carbon and renewable energy economy survey: 2016 final estimates. Statistical bulletin, Office for National Statistics

ONS (2018) Gross domestic product (GDP). Office for National Statistics

Planelles M (2017) La Receta Británica para desengancharse del carbón. In: EL PAIS Madrid, Spain. Accessed 05 Jun 2018

PWC (2017) Is Paris possible? The Low Carbon Economic Index 2017

Quartz (2017) The UK could have changed the way the world fights global warming. Instead it blew $200 million. https://qz.com/972939/the-uk-could-have-changed-the-way-the-world-fights-global-warming-instead-it-blew-200-million/. Accessed 11 Jun 2018

Roberts SH et al (2018) Consequences of selecting technology pathways on cumulative carbon dioxide emissions for the United Kingdom. Appl Energy 228:409–425

Romm J (2016) Climate change, what everyone needs to know. O.U. Press

Schiermeier Q (2014) 'Modest' EU climate targets criticized. Nature. https://www.nature.com/news/modest-eu-climate-targets-criticized-1.14573. Accessed 06 Aug 18

Smith P et al (2015) Biophysical and economic limits to negative CO_2 emissions. Nat Clim Change 6:42

Upham P, Kivimaa P, Virkamäki V (2013) Path dependence and technological expectations in transport policy: the case of Finland and the UK. J Transp Geogr 32:12–22

WRI (2016) The roads to decoupling: 21 Countries are reducing carbon emissions while growing GDP. http://www.wri.org/blog/2016/04/roads-decoupling-21-countries-are-reducing-carbon-emissions-while-growing-gdp. Accessed 06 Jun 2018

Zuluaga D (2017) A post-Brexit framework for electricity markets. Institute of Economic Affairs

Chapter 8
Measuring Enteric Methane Emissions from Individual Ruminant Animals in Their Natural Environment

Matt J. Bell

Abstract Ruminant livestock are an important source of meat, milk, fiber, and labor for humans. The process by which ruminants digest plant material through rumen fermentation into useful product results in the loss of energy in the form of methane gas from consumed organic matter. The animal removes the methane building up in its rumen by repeated eructations of gas through its mouth and nostrils. Ruminant livestock are a notable source of atmospheric methane, with an estimated 17% of global enteric methane emissions from livestock. Historically, enteric methane was seen as an inefficiency in production and wasted dietary energy. This is still the case, but now methane is seen more as a pollutant and potent greenhouse gas. The gold standard method for measuring methane production from individual animals is a respiration chamber, which is used for metabolic studies. This approach to quantifying individual animal emissions has been used in research for over 100 years; however, it is not suitable for monitoring large numbers of animals in their natural environment on commercial farms. In recent years, several more mobile monitoring systems discussed here have been developed for direct measurement of enteric methane emissions from individual animals. Several factors (diet composition, rumen microbial community, and their relationship with morphology and physiology of the host animal) drive enteric methane production in ruminant populations. A reliable method for monitoring individual animal emissions in large populations would allow (1) genetic selection for low emitters, (2) benchmarking of farms, and (3) more accurate national inventory accounting.

Keywords Enteric methane · Measurements · Methods · Normal environment

---

M. J. Bell (✉)
University of Nottingham, Sutton Bonington, Leicestershire LE12 5RD, UK
e-mail: matt.bell@nottingham.ac.uk

© Springer Nature Singapore Pte Ltd. 2019
N. Shurpali et al. (eds.), *Greenhouse Gas Emissions*, Energy, Environment, and Sustainability, https://doi.org/10.1007/978-981-13-3272-2_8

8.1 Enteric Methane Production

Mankind relies on domesticated herbivorous mammals of the Bovidae family (about 3.6 billion worldwide (Hackmann and Spain 2010)), such as ruminants, to produce edible food (e.g., meat and milk), fiber, and labor. Importantly, ruminants are efficient convertors of non-human edible plant material into edible energy and protein. A total 37% of the world's terrestrial land area is grassland and provides a natural and potential source of affordable nutrients for animals if managed sustainably (Suttie et al. 2005).

Ruminants are responsible for an estimated 17% of global enteric methane emissions and 3.3% of total global greenhouse gas emissions from anthropogenic sources (Knapp et al. 2014). Attributes of the ruminant animal and its diet all influence the amount of methane produced (Fig. 8.1). The main drivers of enteric methane production are diet composition, the rumen microbial community, and their symbiotic relationship with the morphology and physiology of the animal's digestive system.

8.1.1 Rumen Microbes and Methane Production

Ruminants have evolved a four-chamber foregut that includes the rumen, which contains bacteria, protozoa, and fungi that ferment plant material with a by-product being the production of metabolic hydrogen and its utilization by methanogenic

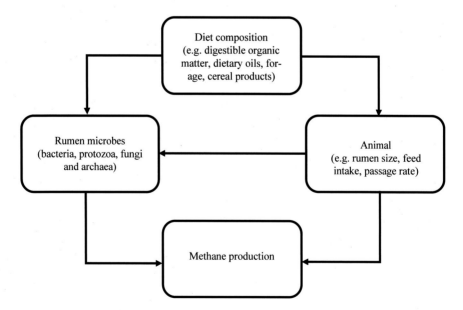

Fig. 8.1 Main drivers of enteric methane production

archaea to produce methane gas. Ruminants lack the enzymes needed to degrade plant polysaccharides, and instead rely on a diverse community of rumen microbes. The addition of cellulase and hemicellulase enzymes to a ruminant's diet may also enhance fiber digestion and productivity (Beauchemin et al. 2008). The rumen bacteria, fungi, and protozoa ferment consumed food to form volatile fatty acids that provide a source of energy for the animal. Eventually, the microbial biomass and some unfermented feed components (such as dietary fats and undigested organic matter) pass into the hindgut further providing a potential source of nutrients. In ruminants, enteric methane is produced predominantly in the rumen (87–93%) rather than the hindgut (Kebreab et al. 2006). Any methane produced in the hindgut is largely (i.e., 90%) absorbed and expired through the lungs, with the remainder being excreted through the rectum (Murray et al. 1976). The loss of methane from the rectum has been estimated at between 1 and 8% (Murray et al. 1976; Johnson et al. 1994; Grainger et al. 2007), with the lower value being associated with sheep and the higher value for dairy cattle.

The process of methane production (Knapp et al. 2014) in the rumen and hindgut is as follows: Glucose equivalents from plant polysaccharides (e.g., cellulose, hemicellulose, pectin, starch, and sucrose) are hydrolyzed by microbial enzymes to form pyruvate (8.1).

$$Glucose \rightarrow 2pyruvate + 4H \tag{8.1}$$

The anaerobic fermentation by bacteria, protozoa, and fungi of pyruvate produces reduced co-factors such as NADH. Reduced co-factors are then re-oxidized (e.g., NADH to NAD) to complete the synthesis of volatile fatty acids; the principle products being acetate, butyrate, and propionate (anions of acetic, butyric, and propionic acids).

$$Pyruvate + H_2O \rightarrow acetate \ (C_2) + CO_2 + 2H \tag{8.2}$$

$$2C_2 + 4H \rightarrow butyrate \ (C_4) + 2H_2O \tag{8.3}$$

The creation of acetate (8.2) and butyrate (8.3) provides a source of metabolic hydrogen. Alternatively, the production of propionate (8.4) can utilize available hydrogen and reduce the potential for methane to be produced.

$$Pyruvate + 4H \rightarrow propionate \ (C_3) + H_2O \tag{8.4}$$

The available metabolic hydrogen is converted to hydrogen gas by hydrogenase-expressing bacteria, and then the hydrogen gas is utilized by methanogens (methanogenesis) to produce methane and water (8.5).

$$4H_2 + CO_2 \rightarrow CH_4 + 2H_2O \tag{8.5}$$

Therefore, the diet of the animal influences the production and balance of volatile fatty acids in the rumen. For example, if the ratio of acetate to propionate was greater than 0.5, then hydrogen will accumulate to be used by methanogens (Johnson and Johnson 1995). A buildup of hydrogen is potentially detrimental to the animal as it can inhibit microbial growth, forage digestion, and production of volatile fatty acids (Moss et al. 2000). While production of methane by methanogens is the main sink for available hydrogen, there are two other lesser but alternative sinks for available hydrogen which are (1) the saturation of unsaturated fatty acids (dehydrogenation) and (2) the production of ammonia from the degradation of amino acids.

8.1.2 Diet Composition and Feed Intake

Diets high in fiber content promote rumen bacteria that produce acetate. Diets containing more rapidly fermented plant carbohydrates such as starch and sugar, promote rumen bacteria that produce propionate. Changes in diet and available substrate result either in a shift in the microbial population or a reduction in fermentation rate. For highly fermentable diets, the production of propionate can exceed the current requirement of the animal and its ability to buffer a change in rumen pH (pH below about 5.5). This leads to the production of lactic acid and a change in the microbial population. The ratio of acetate to propionate varies depending on the relative proportions of different rumen bacteria and due to the animal's diet. The rate of rumen microbial fermentation, and availability of metabolic hydrogen, at any given time determines the production of substrates. While both diet composition, morphology and physiology of the host animal influence the microbial community (Weimer et al. 2010; Guan et al. 2008), diet composition appears to have greater influence than the host animal (Henderson et al. 2018). In ruminants, archaea (majority being methanogens) have been found to be less diverse than rumen bacteria, reflecting the narrow range of substrates that archaea depend upon (Henderson et al. 2018). Furthermore, Henderson et al. (2018) found the archaeal groups of *Methanobrevibacter gottschalkii*, *Methanobrevibacter ruminantium*, *Methanosphaera sp.*, and two Methanomassiliicoccaceae-affiliated groups account for 89% of methanogen communities globally. The diversity of rumen bacteria and interaction with the host animal may explain differences in methane emissions among sheep fed the same diet in a study by Bell et al. (2016). Data from the Rowett Institute in Scotland (2016) showed measured metabolizable energy values for paired sheep fed the same diet and amount of feed were highly correlated (Lin's concordance correlation coefficient = 0.93; Fig. 8.2), but considerable variation existed in methane produced per kilogram dry matter intake between paired sheep on the same diet (Lin's concordance correlation coefficient = 0.22; Fig. 8.3).

The effect of diet, i.e., amount of intake and composition, has been found to account for a large proportion of variation in enteric methane emissions from animals (Mills et al. 2003; Bell and Eckard 2012). It is well recognized that

8 Measuring Enteric Methane Emissions …

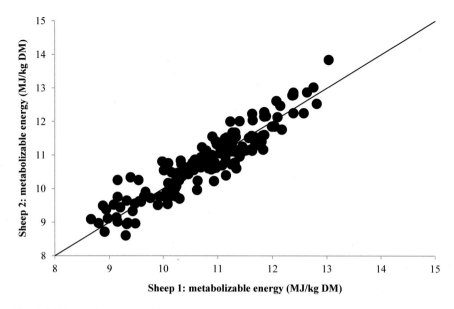

Fig. 8.2 Observed metabolizable energy values (g/kg dry matter (DM)) for paired sheep (n pairs = 144) fed the same diet and amount of feed. A 45° line through the origin is shown (Bell et al. 2016)

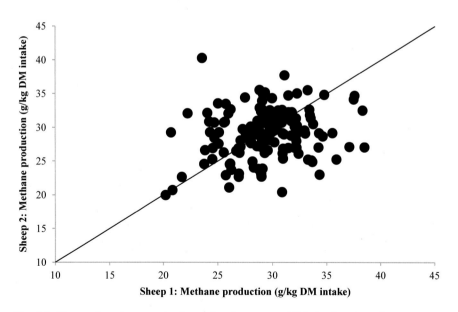

Fig. 8.3 Observed methane production (g/kg dry matter (DM) intake) for paired sheep (n pairs = 144) fed the same diet and amount of feed. A 45° line through the origin is shown (Bell et al. 2016)

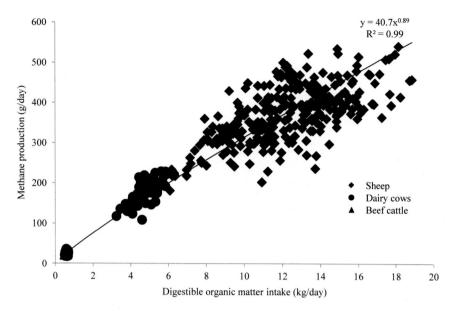

Fig. 8.4 Relationship between digestible organic matter intake and methane production per day for sheep (♦; n = 288), beef cattle (▲; n = 71), and dairy cows (■; n = 284). The line of best-fit across all values and passing through the origin is shown (Bell et al. 2016)

methane production is positively associated with dry matter intake and in particular digestible organic matter intake in ruminant livestock (R^2 = 0.99; Fig. 8.4).

Even with the high association between digestible organic matter intake and methane production seen across ruminant livestock, there is notable variation in emissions at a given level of intake, which is particularly noticeable for dairy cows (Fig. 8.4) and also seen in sheep (Fig. 8.3). Across cattle and sheep fed diets encompassing a wide range of nutrient concentrations (i.e., 235–649 g NDF/kg dry matter, 92–251 g crude protein/kg dry matter, 17–64 g ether extract/kg dry matter, and 9 to 14 MJ metabolizable energy/kg dry matter) and methane emissions (14–40 g/kg dry matter), Bell et al. (2016) found digestible organic matter, oil (ether extract), and feeding level (metabolizable energy intake expressed as multiples of maintenance requirement) as the important explanatory variables describing methane per kilogram of dry matter intake (8.6).

$$CH_4(g/kg\ DM\ intake) = 0.046\,(s.e.\,0.001) \times \text{digestible organic matter} \\ - 0.113\,(s.e.\,0.023) \times \text{oil (both g/kg DM)} \\ - 2.47\,(s.e.\,0.29) \times (\text{feeding level} - 1) \quad (8.6)$$

As expected, there is a positive response in methane produced to per unit dry matter intake to increasing digestible organic matter. The positive response to increasing digestible organic matter can be reduced by increasing dietary contents

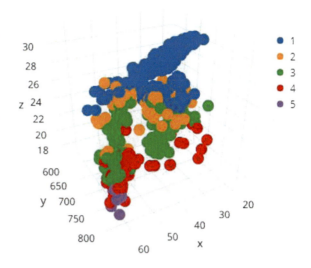

Fig. 8.5 Effect of diet contents of digestible organic matter (y), oil (x) and feeding level from 1 to 5 (metabolizable energy intake expressed as multiples of maintenance requirement) on methane production (z) by cattle and sheep (n = 643) per kilogram dry matter intake (y values adjusted for the random effect of experiment)

of oil and/or increasing feeding level (Fig. 8.5). Due to their chemical composition, individual feed ingredients can vary considerably in their methanogenic effect, with distiller's grains resulting in 3.8% of gross energy intake losses as methane and peas 12.8% (Giger-Reverdin and Sauvant 2010). Diets that encourage a higher rate of fermentation increase the passage rate of food through the rumen and potential level of feed intake, therefore reducing methane losses per unit intake (Johnson and Johnson 1995; Blaxter and Clapperton 1965). While increased intake of less digestible feeds such as forage and fiber can result in increased acetate and methane production, there appears little effect on methane production per dry matter intake (Blaxter and Clapperton 1965). Whereas an increase in more digestible feeds, such as cereal products in the diet, gives rise to elevated levels of propionate resulting in a curvilinear reduction in methane losses per dry matter intake (Holter and Young 1992; Reynolds et al. 2011). Easily digestible diets can lose as little as 2–3% of gross energy intake as methane, whereas less digestible diets with often more than 50% forage content would be associated with greater than 6% of gross energy intake loss (Czeskawski 1988). Also, increasing the dietary oil content in diets inhibits fiber digestion (Ramin and Huhtanen 2013; Moate et al. 2015) and encourages post-ruminal digestion, particularly in the small intestine, which is energetically more efficient with less methane losses than in the rumen.

Animals with the highest feeding levels (4 and 5 times maintenance requirements) and fed diets with high oil content have the lowest emissions per unit intake (Fig. 8.5). Low enteric methane losses per unit intake appear possible by mechanisms that promote the passage of organic matter to post-rumen digestion and reduce rumen fermentation by high intakes of digestible feed and addition of dietary oil.

8.2 Direct Measurement of Individual Animal Methane Emissions

Several studies have compared different approaches for measuring methane emissions from individual animals (Kebreab et al. 2006; Johnson and Johnson 1995; Hammond et al. 2016; Storm et al. 2012). Studies with a respiration calorimeter (chamber) investigating the metabolic efficiency of cattle and sheep fed different diet treatments provide a measure of methane output and assessment of variation among animals. However, such an approach is not applicable for population studies on commercial farms. The worldwide interest in measuring methane emissions from individual animals appears justified given the considerable variation seen among animals fed the same diet (Fig. 8.3), and the benefit this would bring to advancing our ability to monitor this anthropogenic source of emissions. This has led in recent years to the development of approaches that take repeated 'spot' measurements of methane from the breath of animals in their natural environment.

8.2.1 Whole Animal Emissions

Historically, most studies assessing methane emissions and energy efficiency of ruminant livestock have been done using a respiration chamber (Yan et al. 2009, 2010). The respiration chamber is recognized as the gold standard method for measuring whole animal methane losses (i.e., mouth, nostril, and flatulence; Fig. 8.6).

This method involves fresh air flow in and extracted by a pump or fan out of the chamber. The air concentrations (i.e., oxygen, carbon dioxide, hydrogen, and methane) in the incoming and outgoing air are measured at intervals using an

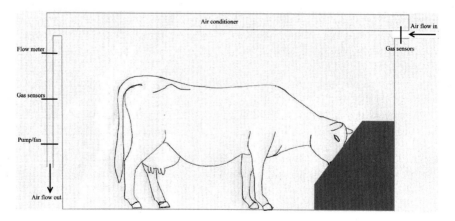

Fig. 8.6 Illustration of a respiration chamber for measuring whole animal gaseous emissions

arrangement of gas sensors to determine the gas emission rate produced by the animal. The gas emission rates are multiplied by airflow to finally derive daily gas production. Chamber temperature, humidity, and the mixing of air are often controlled using an air conditioner.

Housing individual animals in a respiration chamber for usually three days (final two days being used to derive animal gas production) is impractical for large-scale measurements of methane from individual animals. Also, housing an animal in a chamber can affect individual animals differently, and potentially result in depression of appetite (Murray et al. 1999), which is less of an issue for comparing feed treatments in whole animal metabolic studies than differences among animals. The impact on animal behavior can be minimized by ensuring visual contact with other animals and familiarity with the environment (Storm et al. 2012).

A more mobile and smaller chamber has been developed for sheep, a portable accumulation chamber (PAC), which measures gas emissions from individual chambers for up to 1 h (Jonker et al. 2018). The results appear less repeatable than respiration chamber measurements, presumably partly explained by the contrasting environments when sampling; however, the small chamber can be used on commercial farms for short periods and with grazing systems. Other sampling methods measuring whole animal emissions have used an enclosed barn (Johnson et al. 2002), polythene tunnel (Murray et al. 1999), or simply in the field using a tracer gas (Griffith et al. 2008). These approaches require careful monitoring of the sampling environment, which makes replicating these techniques consistently on commercial farms difficult.

More invasive methods used in research and not appropriate for commercial farm use involve injecting radioactively labeled methane (isotope dilution technique) (Murray et al. 1976; France et al. 1993) or ethane (Moate et al. 1997) into the rumen fluid and gas sampled by cannula or within an enclosure such as a chamber.

8.2.2 Breath Sampling

Measurement methods that try to integrate into the natural environment of the animal have been developed that measure solely methane produced from the mouth and nostrils of the animal (since this represents the majority of the animal's emissions). This approach has been found to correlate well with respiration chamber measurements (Grainger et al. 2007; Garnsworthy et al. 2012). However, due to the often-higher variability observed with this approach, the number of animals and days needed to assess treatment effects using breath sampling methods are greater than when using respiration chambers. Typically, a minimum of 5–7 days of 'spot' measurements are needed. The duration of sampling needed to obtain repeatable measurement that allows assessment of within-cow, between-cow, diet, and temporal effects is dependent on the frequency of spot measurements, which can be influenced by visits to the sampling location, and the ability to account for potential sources of error (Cottle et al. 2015).

One such approach is the sulfur hexafluoride (SF_6) tracer method (Fig. 8.7) which involves collecting breath samples continuously into an evacuated canister over a period of several hours within a day and for 5–7 days (Johnson et al. 1994). The air inlet of a capillary tube is held close to the nostril of the animal by a head halter. A permeation tube containing a known amount of the inert gas SF_6 is placed in the rumen of the animal and continuously releases the gas over the sampling period. Prior to placing the permeation tube in the animal's rumen, the release rate of SF_6 from each tube is determined by placing the tube in a water bath at 39 °C and routinely weighing the tube until an accurate loss rate is obtained. The ratio of concentrations for methane and SF_6 collected in the canister on the animal, and analyzed using gas chromatography, along with the release rate of SF_6 gas (QSF_6) from the permeation tube are used to derive the methane emission rate (QCH_4) and daily methane production (8.7).

$$QCH_4 = QSF_6 \times [CH_4]/[SF_6] \qquad (8.7)$$

While the use of the SF_6 technique shows good agreement with methane emissions measured from the same cows in a respiration chamber, the approach appears to produce more variable results (Grainger et al. 2007; Pinares-Patiño et al. 2011). Some of this variability can be attributed to the invasive nature of the equipment, consistency of release/sampling of the SF_6 gas, and influence of background gas concentrations. The use of a tracer gas is not always permitted in every country. The method may also be more suited to animals fed a high-forage

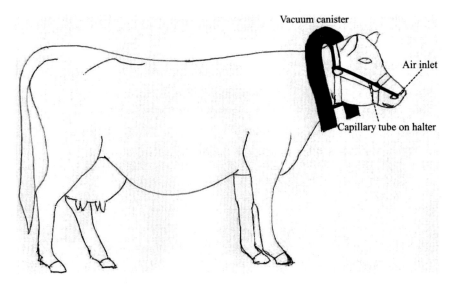

Fig. 8.7 Illustration of the sulfur hexafluoride (SF_6) tracer method for measuring methane emissions from the nostril of the animal

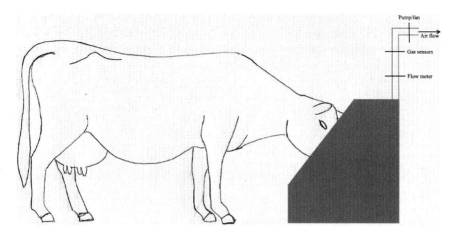

Fig. 8.8 Illustration of the sniffer method for measuring methane emissions from the mouth and nostrils of the animal at a feed bin

diet and not with diets that result in greater post-ruminal digestion (McGinn et al. 2006).

Other methods have been developed to sample methane emissions from solely the breath of an animal using a head box (Kelly et al. 1994), mask (Liang et al. 1989), at a feed bin (Garnsworthy et al. 2012; Huhtanen et al. 2015), or with a laser gun methane detector pointed at the muzzle of the animal (Chagunda et al. 2009) (Fig. 8.8). The use of a head box, mask, and laser gun approach may require the animal to be restrained and limit the animal's ability to drink and eat and function normally.

These methods involve differing levels of complexity (i.e., flow meters, tracer gas, attachments, proximity sensor, and filters) and use frequent 'spot' measurements within a day (rather than continuously over 24 h as with the chamber) to determine methane production. The regular sampling of gas within a day needs to ensure that it accounts for the head position of the animal in close proximity to the sampling tube (i.e., when a peak in gas concentration is observed, Fig. 8.9) and the diurnal pattern for methane (Fig. 8.10). The location of the animal's head to the gas sampling tube can be determined using a proximity sensor (Huhtanen et al. 2015) or filtering the data for eructation peaks of methane (Garnsworthy et al. 2012). Figure 8.9 shows eructation peaks for two cows measured during milking at a feed bin, with gas concentration measured every second and with an air flow rate of one liter per minute. Both Cow A and Cow B milked for a similar length of time and consumed a similar amount of a commercial ration (50% forage in the dry matter) during the day (19.7 and 19.1 kg dry matter intake, respectively). However, Cow A had a higher eructation rate of 1.3 per minute (mean peak concentration of 728 ppm) compared to Cow B of 1.0 eructation per minute (mean peak concentration of 847 ppm). These differences in mean concentration and frequency of eructations can be combined to derive individual animal methane emission rate.

Bell et al. (2014a) measured the emission rate of 1,964 dairy cows on commercial farms in the UK and found an average individual cow emission rate of 2.9 mg/min (ranging from 0.6 to 4.8 mg/min). This equates to an average of 418 g/day per cow and a range of 286 to 526 g/day using the equation by Garnsworthy et al. (2012) (methane (g/day) = 252 + (57.2 × emission rate in mg/min)), which links on-farm and chamber measurements. This range of values is similar to the range reported for dairy cows of 220 to 480 g/day by Grainger et al. (2007). With a population of about 1.8 million dairy cows in the UK, this would amount to approximately 275 thousand tonnes of methane produced each year.

Within periods of 'spot' measurements, the frequency of eructations and gas concentration of eructations vary among cows (Fig. 8.9). Also, methane emissions over a 24-hour day are characterized by a diurnal pattern (Crompton et al. 2011; Manafiazar et al. 2017; Bell et al. 2018), with a peak in emissions after feeding being followed by a gradual decline until the next consumption of feed. The average diurnal pattern across a group of animals often shows a peak in methane production soon after the feed is allocated due to this activity stimulating most animals to feed (Fig. 8.10). However, in reality, there is considerable variation in diurnal patterns among animals due to the time when animals choose to eat (Bell et al. 2018).

The diurnal pattern is affected by feed allowance and feeding frequency (Crompton et al. 2011) and does not appear to change over time or with a change in diet (Bell et al. 2018). The frequent 'spot' sampling of breath methane emissions has come about due to the need to measure methane from commercial animals. Methods that are more mobile, non-invasive to the animal and can fit into the

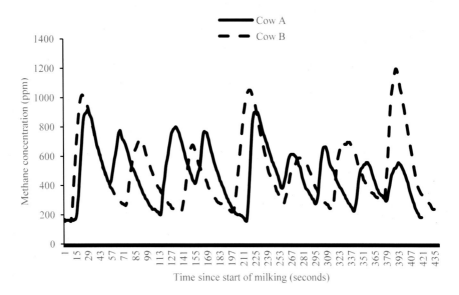

Fig. 8.9 Methane measured continuously during milking from a feed bin showing a profile for Cow A and Cow B

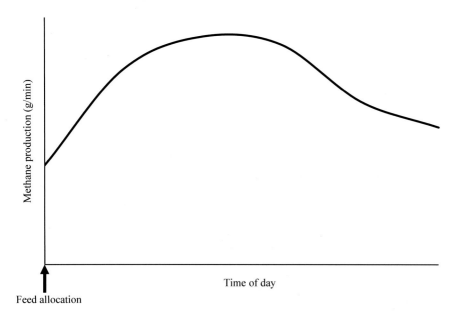

Fig. 8.10 Average diurnal pattern of methane emissions during a 24-hour day for cows allocated food once per day

animal's natural environment are of great interest, but bring challenges in application and data processing. Taking 'spot' measurements of methane (expressed in various units of concentration, emission rate, ratio with carbon dioxide, or grams/day) has been found to be a repeatable measure (Huhtanen et al. 2015; Bell et al. 2014b); however, to be a reliable measure, the data requires processing to account for sources of error such as cow-head position (Huhtanen et al. 2015), number and timing of measurements (Hammond et al. 2016; Cottle et al. 2015), and potential changes in the sampling environment. Overall, this approach and development can provide useful insight into quantifying methane emissions on commercial farms as illustrated and explore sources of variation in large populations of animals.

References

Beauchemin KA, Kreuzer M, O'Mara F, McAllister TA (2008) Nutritional management for enteric methane abatement: a review. Aust J Exp Agric 48:21–27

Bell MJ, Eckard RJ (2012) Reducing enteric methane losses from ruminant livestock—Its measurement, prediction and the influence of diet. In: Javed K (ed) Livestock production. In Tech Publishing, Rijeka, Croatia, pp 135–150

Bell MJ, Potterton S, Craigon J, Saunders N, Wilcox R, Hunter M, Goodman JR, Garnsworthy PC (2014a) Variation in enteric methane emissions among cows on commercial dairy farms. Animal 8:1540–1546

Bell MJ, Saunders N, Wilcox R, Homer E, Goodman JR, Craigon J, Garnsworthy PC (2014b) Methane emissions among individual dairy cows during milking quantified by eructation peaks or ratio with carbon dioxide. J Dairy Sci 97:6536–6546

Bell MJ, Eckard R, Moate PJ, Yan T (2016) Modelling the effect of diet composition on enteric methane emissions across sheep, beef cattle and dairy cows. Animals 6:54

Bell MJ, Craigon J, Saunders N, Goodman JR, Garnsworthy PC (2018) Does the diurnal pattern of enteric methane emissions from dairy cows change over time? Animal 1–6. https://doi.org/10.1017/s1751731118000228

Blaxter KL, Clapperton JL (1965) Prediction of the amount of methane produced by ruminants. Br J Nutr 19:511–522

Chagunda MGG, Ross D, Roberts DJ (2009) On the use of a laser methane detector in dairy cows. Comput Electron Agric 68:157–160

Cottle DJ, Velazco J, Hegarty RS, Mayer DG (2015) Estimating daily methane production in individual cattle with irregular feed intake patterns from short-term methane emission measurements. Animal 9:1949–1957

Crompton LA, Mills JAN, Reynolds CK, France J (2011) Fluctuations in methane emission in response to feeding pattern in lactating dairy cows. In: Modelling nutrient digestion and utilization in farm animals. In: Sauvant D, Van Milgen J, Faverdin P, Friggens N (eds) Wageningen Academic Publishers, Wageningen, the Netherlands, pp 176–180

Czeskawski JW (1988) An introduction to Rumen studies. Pergamon Press, Oxford, UK

France J, Beever DE, Siddons RC (1993) Compartmental schemes for estimating methanogenesis in ruminants from isotope dilution data. J Theor Biol 164:207–218

Garnsworthy PC, Craigon J, Hernandez-Medrano JH, Saunders N (2012) On-farm methane measurements during milking correlate with total methane production by individual dairy cows. J Dairy Sci 95:3166–3180

Giger-Reverdin S, Sauvant D (2010) Methane production in sheep in relation to concentrate feed composition from bibliographic data. Cahiers Opt Méditerranéennes 52:43–46

Grainger C, Clarke T, McGinn SM, Auldist MJ, Beauchemin KA, Hannah MC, Waghorn GC, Clark H, Eckard RJ (2007) Methane emissions from dairy cows measured using the sulphur hexafluoride (SF_6) tracer and chamber techniques. J Dairy Sci 90:2755–2766

Griffith DWT, Glenn R, Bryant DH, Reisinger AR (2008) Methane emissions from free-ranging cattle: comparison of tracer and integrated horizontal flux techniques. J Environ Qual 37:582–591

Guan LL, Nkrumah JD, Basarab JA, Moore SS (2008) Linkage of microbial ecology to phenotype: Correlation of rumen microbial ecology to cattle's feed efficiency. FEMS Microbiol Lett 288:85–91

Hackmann TJ, Spain JN (2010) Invited review: ruminant ecology and evolution: perspectives useful to ruminant livestock research and production. J Dairy Sci 93:1320–1334

Hammond KJ, Crompton LA, Bannink A, Dijkstra J, Yáñez-Ruiz DR, O'Kiely P, Kebreab E, Eugenè MA, Yu Z, Shingfield KJ, Schwarm A, Hristov AN, Reynolds CK (2016) Review of current in vivo measurement techniques for quantifying enteric methane emission from ruminants. Anim Feed Sci Technol 219:13–30

Henderson G, Cox F, Ganesh S, Jonker A, Young W, Janssen PH (2018) Rumen microbial community composition varies with diet and host, but a core microbiome is found across a wide geographical range. Scientif Rep 5:14567

Holter JB, Young AJ (1992) Methane prediction in dry and lactating Holstein cows. J Dairy Sci 75:2165–2175

Huhtanen P, Cabezas-Garcia EH, Utsumi S, Zimmerman S (2015) Comparison of methods to determine methane emissions from dairy cows in farm conditions. J Dairy Sci 98:3394–3409

Johnson KA, Johnson DE (1995) Methane emissions from cattle. J Anim Sci 73:2483–2492

Johnson KA, Huyler MT, Westberg HH, Lamb BK, Zimmerman P (1994) Measurement of methane emissions from ruminant livestock using a SF6 tracer technique. Environ Sci Technol 28:359–362

Johnson KA, Kincaid RL, Westberg HH, Gaskins CT, Lamb BK, Cronrath JD (2002) The effect of oilseeds in diets of lactating cows on milk production and methane emissions. J Dairy Sci 85:1509–1515

Jonker A, Hickey SM, Rowe SJ, Janssen PH, Shackell GH, Elmes S, Bain WE, Wing J, Greer GJ, Bryson B, MacLean S, Dodds KG, Pinares-Patiño CS, Young EA, Knowler K, Pickering NK, McEwan JC (2018) Genetic parameters of methane emissions determined using portable accumulation chambers in lambs and ewes grazing pasture and genetic correlations with emissions determined in respiration chambers. J Anim Sci https://doi.org/10.1093/jas/sky187

Kebreab E, Clark K, Wagner-Riddle C, France J (2006) Methane and nitrous oxide emissions from Canadian animal agriculture: a review. Canadian J Anim Sci 86:135–158

Kelly JM, Kerrigan B, Milligan LP, McBride BW (1994) Development of a mobile, open circuit indirect calorimetry system. Canad J Anim Sci 74:65–72

Knapp JR, Laur GL, Vadas P, Weiss WP, Tricarico JM (2014) Invited review: enteric methane in dairy cattle production: quantifying the opportunities and impact of reducing emissions. J Dairy Sci 97:3231–3261

Liang JB, Terada F, Hamaguchi I (1989) Efficacy of using the face mask technique for the estimation of daily heat production of cattle. In: Van Der Honing Y, Close WH (ed) Energy metabolism of farm animals. Pudoc, Waginingen, The Netherlands (1989)

Manafiazar G, Zimmerman S, Basarab JA (2017) Repeatability and variability of short-term spot measurement of methane and carbon dioxide emissions from beef cattle using GreenFeed emissions monitoring system. Canad J Anim Sci 97:118–126

McGinn SM, Beauchemin KA, Iwaasa AD, McAllister TA (2006) Assessment of the sulfur hexafluoride (SF_6) tracer technique for measuring enteric methane emissions from cattle. J Environ Qual 35:1686–1691

Mills JAN, Kebreab E, Yates CM, Crompton LA, Cammell SB, Dhanoa MS, Agnew RE, France J (2003) Alternative approaches to predicting methane emissions from dairy cows. J Anim Sci 81:3141–3150

Moate PJ, Clarke T, Davies LH, Laby RH (1997) Rumen gases and load in grazing dairy cows. J Agric Sci 129:459–469

Moate P, Deighton MH, Richard S, Williams O, Pryce JE, Hayes BJ, Jacobs JL, Eckard RJ, Hannah MC, Wales WJ (2015) Reducing the carbon footprint of Australian milk production by mitigation of enteric methane emissions. Anim Product Sci 56:1017–1034

Moss AR, Jouany J-P, Newbold J (2000) Methane production by ruminants: its contribution to global warming. Annal de Zootechnie 49:231–253

Murray RM, Bryant AM, Leng RA (1976) Rates of production of methane in the rumen and large intestines of sheep. Br J Nutr 36:1–14

Murray PJ, Moss A, Lockyer DR, Jarvis SC (1999) A comparison of systems for measuring methane emissions from sheep. J Agric Sci 133:439–444

Pinares-Patiño CS, Lassey KR, Martin RJ, Molano G, Fernandez M, MacLean S, Sandoval E, Luo D, Clark H (2011) Assessment of the sulphur hexafluoride (SF6) tracer technique using respiration chambers for estimation of methane emissions from sheep. Anim Feed Sci Technol 166–167:201–209

Ramin M, Huhtanen P (2013) Development of equations for predicting methane emissions from ruminants. J Dairy Sci 96:1–18

Reynolds CK, Crompton LA, Mills JAN (2011) Improving the efficiency of energy utilisation in cattle. Anim Product Sci 51:6–12

Storm IMLD, Hellwing ALF, Nielsen NI, Madsen J (2012) Methods for measuring and estimating methane emission from ruminants. Animals 2:160–183

Suttie JM, Reynolds SG, Batello C (2005) Grasslands of the world. Food and Agriculture Organization of the United Nations, Rome

Weimer PJ, Stevenson DM, Mantovani HC, Man SLC (2010) Host specificity of the ruminal bacterial community in the dairy cow following near-total exchange of ruminal contents. J Dairy Sci 93:5902–5912

Yan T, Porter MG, Mayne CS (2009) Prediction of methane emission from beef cattle using data measured in indirect open-circuit respiration calorimeters. Animal 3:1455–1462

Yan T, Mayne CS, Gordon FG, Porter MG, Agnew RE, Patterson DC, Ferris CP, Kilpatrick DJ (2010) Mitigation of enteric methane emissions through improving efficiency of energy utilization and productivity in lactating dairy cows. J Dairy Sci 93:2630–2638

Chapter 9
Crop Residue Burning: Effects on Environment

Ritu Mathur and V. K. Srivastava

Abstract The majority of the Indian population is dependent on agriculture for livelihood. While agricultural activities are a source of income for around 70% of the Indian population and cater to the food and nutritional needs of the vast nation, the practice of burning residue of crops has devastating impacts on the environment. Burning the biomass residue is the easiest and most time-efficient mode of ensuring that the field is ready for the next cropping season in time. Needless to say that this practice is not environmentally sustainable. The emission of greenhouse gases like carbon dioxide, methane and nitrous oxide is one such consequence which has resulted in multiple environmentally degrading phenomena such as air pollution, global warming, smog and climate change. In order to cater to the food requirements of the 1.2 billion Indians and continue the dependence of the majority of them on agriculture, a balance needs to be struck so that the activity is sustainable and does not cause irreparable harm to the planet. A few such solutions are using the crop biomass as fodder for livestock, utilising the residue for power generation and converting the waste material to natural fuels. Not only will this reduce the plight of the already fragile and sensitive environment, but it would also lead to an increase in the income of the farmers. This would also be an effective waste management solution.

Keywords Greenhouse gases · Crop residue · Global warming
Air pollution · Waste management

R. Mathur (✉)
Government R.R. Autonomous College, Alwar, Rajasthan, India
e-mail: ritu.chem@gmail.com

V. K. Srivastava
Sankalchand Patel University, Visnagar, Gujarat, India
e-mail: drvks9@gmail.com

© Springer Nature Singapore Pte Ltd. 2019
N. Shurpali et al. (eds.), *Greenhouse Gas Emissions*, Energy, Environment,
and Sustainability, https://doi.org/10.1007/978-981-13-3272-2_9

9.1 Introduction

Living in an era of rapidly escalating environmental crisis, we need to sensitise the world about its hazards and the various possible solutions.

Greenhouse gases (hereinafter, GHGs) like carbon dioxide, methane and nitrous oxide interact with the sun's energy, trap the heat in the Earth's atmosphere and cause the greenhouse effect. Human activities like denudation of forests, combustion of fossil fuels, and natural factors like volcanoes emit GHGs and result in global warming. Adverse effects of global warming can be witnessed upon climate, plants, human health, wildlife, etc. (https://www.ipcc.ch/ipccreports/tar/wg1/pdf/TAR-01.PDF).

Indian economy is mainly governed by the agricultural sector. More than half of the Indian population is dependent on agriculture for livelihood. However, the productivity of this sector is extremely low, as evidenced by its contribution of a mere 16% to the gross domestic product (hereinafter, GDP) of India (http://www.indiaenvironmentportal.org.in/files/file/economic%20survey%202017-18%20-%20vol.1.pdf). India follows a wheat–rice cropping pattern as both crops are cultivated on the same land alternatively. The time gap between harvesting of rice crop and sowing of wheat crop is only 15–20 days (Hobbs and Morris 1996). Although wheat straws (residue) are largely removed as cattle fodder, the rice or paddy straw (residue) poses a challenge for its disposal. Removal of the residue requires a large amount of labour, and the time window for preparing the farm for the next crop (wheat) is quite small. The easiest and the most economical process, which also requires little labour, is burning the residue in the field itself. This process is known as Agriculture Residue Crop Burning (ARCB). GHGs, particulate matters and air pollutants are emitted due to ARCB, which in turn are responsible for degraded atmosphere, nutrient loss, human health hazards, smog and climate change.

The need of the hour is to channelise this waste into productive purposes, by converting it into natural fuels, using residues as fodder or for power generation. This would ensure sustenance for both the present and the future generations. The present chapter has been divided into four substantive sections. The first section delves into the meaning of the greenhouse effect, along with causes and consequences of global warming. The subsequent section discusses the dependence of the Indian economy on the agriculture sector and also introduces the practice of ARCB. Emissions from crop residue burning and the gases emitted are discussed in the section after. Before concluding, the author deliberates upon possible ideas to facilitate crop residue management, in the last substantive section.

9.2 Indian Economy and Agriculture

The agricultural sector plays a vital role in the Indian economy, and over 49% of the workforce of India is engaged in agricultural activities for sustenance and livelihood. However, despite more than half the working population of the nation being dependent on agriculture, the sectoral contribution to the gross domestic product (GDP) of the country amounts to only about 16% of the total annually (http://www.indiaenvironmentportal.org.in/files/file/economic%20survey%202017-18%20-%20vol.1.pdf). Both owners and cultivators of the land, i.e. tenants and agricultural labourers, form a part of this sector which derives its livelihood from the primary activity of agriculture in India (Deshpande 2017). The dependence of the Indian economy on the agricultural sector can be gauged not only from the fact that nearly half the workforce of the country is dependent on this sector for their livelihood, but also that the whopping 135.42 crores of the Indian population rely on agriculture for its daily food intake.

The prevalent form of agriculture in India is following the wheat–rice cropping pattern, where generally the two crops are cultivated on the same plot in rotation, throughout the year. Vast areas of agricultural lands in major states such as Punjab, Haryana, Himachal Pradesh, Uttar Pradesh, Bihar and Madhya Pradesh witness such a cropping pattern (Hobbs and Morris 1996).

Indian agriculture is faced with the problem of fragmentation of landholding. The same piece of ancestral agricultural land being passed down generation after generation leads to the land being divided among the numerous heirs. The size of the fragmented land, and the continuous reuse of the same, leads to a decline in the fertility of the land, thereby negatively impacting the produce. Given the small size of the land, employing intensive machines to increase efficiency is not a viable option.

Manual harvesting makes use of sickles to manually cut the crop close to the ground, and the remains of the crop are then used as fodder for cattle or used for packaging. However, in case of mechanised farming, nearly 50–60 cm of crop stubble and nearly 40–50 cm of straw are generated and this poses the challenge of disposal of this residue. It is extremely difficult to dispose off especially paddy straw, and this is not a viable option for use as fodder. Farmers then resort to burning of the residue as the quickest mode of disposal and preparation of the field for the next cropping season while also enhancing the fertility of the land (Damodaran 2017). When better quality residue can be used as fodder for their animals, farmers prefer to just dispose off the inferior quality fodder by burning it. If the issue of disposing the crop residue is not taken care of, on a very urgent basis it might lead to serious economic disequilibrium for the generations to come.

Needless to say, the problem of crop residue burning poses serious threats to the stability and the sustainability of the environment. Some of these consequences would be discussed in the subsequent section.

9.3 Emissions from Agricultural Residue Crop Burning

India is an agrarian economy. With the crops, the residue is also an integral part of agricultural production. In the times to come, agricultural production is bound to rise because of the increase in population and so would be the rise in crop residue. Simply put, the waste material generated after the cultivation and harvesting of crops can be understood as crop residue. As discussed in the previous sections, with the increased dependence of the nation on agricultural activities, emerges the problem of crop residue management. In vast tracts of agricultural land, especially in the Northern India, burning of the crop residue is seen as the quickest and the most cost-efficient mode of getting rid of the crop residue (Technology Information Forecasting and Assessment Council 1991).

Biomass burning is the burning of living or dead organic matter which can be initiated by the nature or by the human beings. Forest fires and burning of savannahs (grass plain in tropical or subtropical regions) are natural phenomena which lead to air pollution, whereas ARCB is a human-initiated process.

The main crops may be classified into four categories, namely cereals (rice, wheat, maze, jowar, bajra, ragi, small millets, etc.), oil seeds (groundnut, mustard, rapeseeds, etc.), fibres (cotton, jute, etc.) and sugar cane. The residue of these crops is generated in large quantities in the form of cereal straws, woody stalks and groundnut leaves.

While traditionally this residue was used as fodder, or as packaging material, or to thatch roofs, or as a fuel, these traditional uses have drastically reduced in the current times. The residue of maize and millets is quite hard and cannot, therefore, be used as fodder for animals. The residue of sugar cane and mustard too is mostly burnt, but in some cases is also used as a fuel (Meshram 2002). Residue of groundnut, to some extent, is also used as a fuel (Meshram 2002).

The residue of paddy is generally not considered as fodder due to the presence of high amounts of silica in the same. In order to prepare the field for the wheat crop quickly, the stubble is burnt by the farmers (https://visibleearth.nasa.gov/view.php?id=84680). However, treatment of the crop stubble with urea and moisture can make it fit for consumption by animals as fodder, by increasing the nutritional value and the digestibility of the stubble (Badve 1991).

Haryana, Punjab, Madhya Pradesh, Uttar Pradesh and Maharashtra account for more than half of the crop residue burnt. As per the estimates by the Government of India, over 500 million tonnes of crop residue are produced annually (https://mnre.gov.in/file-manager/annual-report/2016-2017/EN/pdf/3.pdf). The crop residue generation is not, however, uniform across the nation, and there exists a wide variation in the quantity of surplus produced by the various regions in India. The amount of crop residue generated from each state depends on the productivity of the land, the climate conditions, the nature of crops cultivated and the intensity of farming (https://mnre.gov.in/file-manager/annual-report/2016-2017/EN/pdf/3.pdf).

9 Crop Residue Burning: Effects on Environment 131

Jain et al., Aerosol and Air Quality Research, 14: 422–430, 2014

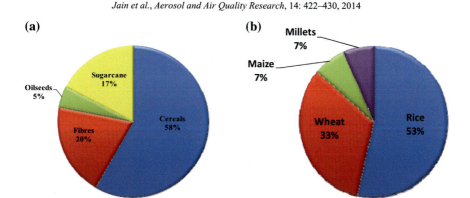

Fig. 9.1 **a** Residue generation from crops; **b** Residue generation from cereal crops. *Source* Jain et al., Aerosol and Air Quality Research, 14: 422–430, 2014

Uttar Pradesh has been recorded to have produced the highest amount of crop residue, followed by Punjab and Maharashtra. These three states produced between 45 and 60 tonnes each of crop residue in the year 2014–15 (Devi et al. 2017, p. 487). The majority of crop residue is produced by cereals, sugar cane, pulses, oilseeds and fibres. Cereals like millets, wheat, rice and maize, being the highest contributor to the pool, contribute nearly 70% of the total crop residue produced, whereas fibre crops contribute nearly 13% of the total crop residue produced (Devi et al. 2017, p. 487) (Fig. 9.1).

According to data compiled by the Ministry of Statistics and Program Implementation (MOSPI), Government of India, there is a gradual increase in the annual crop residue generation in India (1950–2015) which is represented in Fig. 9.2. The crop residue produced per year in 2015 was over 500 million tonnes

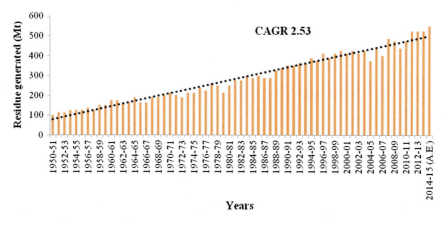

Fig. 9.2 Trend of crop generation in India. *Source* Ministry of Statistics and Program Implementation, 2013–14

(Mt). Around the year 1950, its amount was approximately 80 million tonnes. There has been an increase of almost 575% during this period. Thus, the crop residue generation per year has been increased exponentially.

The impact of burning of the crop residue can be felt not just on the natural environment, but also on human and animal health. The burning also adversely impacts the quality of soil, due to loss of nutrients.

Burning of crop residue has an adverse impact on the environment. The Earth's atmosphere has the assimilative capacity to absorb the pollution generated; however, when activities like crop residue burning generate pollution which is more than what the atmosphere can absorb, the balance in the environment is disturbed, thereby negatively impacting life on the planet (Kumar et al. 2015, p. 13).

The burning of crop residue results in the emission of a multitude of particulate matter and toxic gases such as carbon monoxide, carbon dioxide, methane, dinitrogen monoxide, nitrogen dioxide, sulphur dioxide, volatile organic compounds (VOC), semi-volatile organic compounds (SVOC) and CFCs (Gupta et al. 2004). It also leads to the emission of toxins such as polychlorinated dibenzofurans, polycyclic aromatic hydrocarbons (PAHs) and polychlorinated dibenzo-p-dioxins (Gadde et al. 2009). These potential carcinogens release carbon dioxide in the environment, leading to the depletion of the already diminishing ozone layer, adding to the greenhouse effect (Gadde et al. 2009).

The release of carbon dioxide and carbon monoxide leads to the generation of toxic haemoglobin in the blood supply of animals. This drastically impacts the yield of milk-producing animals and also leads to detriment the health of the humans consuming such milk and milk products. Young children and pregnant women are among the worst sufferers (Singh et al. 2008).

Further, health impacts on human beings include serious respiratory aggravations such as bronchitis, asthma, cough and various eye and skin diseases. The soot generated impacts vision and the fine particulate matter, generated along with the emissions of gases, has been traced to be a cause of acute lung and heart diseases and can also cause early deaths of people suffering from the same (Kumar et al. 2015, p. 24). Asthma and aggravated bronchial attacks are triggered by the inhalation of particulate matter lower than PM 2.5 µg (Kumar et al. 2015, p. 24).

The burning of crop residue is also detrimental to the small flora and fauna and the trees in the vicinity. In addition, the ash left behind having good absorbing properties can lead to the disastrous outcome of the absorption of weedicides. Removal of such ash, however, also leads to the removal of various organic materials which are present naturally in the soil (Gupta et al. 2004). Further, the increase in the temperature of soil caused due to burning leads to the depletion of essential nutrients such as nitrogen, sulphur, potassium, phosphorus and carbon (Gupta et al. 2004).

The process of stubble burning is irrational and unmindful, but as it is a quick and easy way to get rid of the same, it has, thus, become a global phenomenon.

As per studies, burning of the residues from paddy contributed to the emission of toxic gases containing carbon, in the form of carbon dioxide (70%), carbon monoxide (7%) and methane (0.66%). Further, the burning of straw led to the emission of nitrous oxide (20%), nitrous dioxide (2.1%) and sulphur-oxides (17%) (Karlen et al. 1994). Another study shows that ACRB contributed to the emission of 91 Tg per year of carbon dioxide, 4.1 Tg of carbon monoxide per year, 0.6 and 0.1 Tg, respectively, per year of methane and nitrous oxides and 1.2 Tg per year of the total particulate matter (Yevich and Logan 2003).

The burning of agricultural residue not only contributes to gaseous and particulate emissions, but also results in the residue being denuded of nearly all of their nutrients. As per a study, the burning resulted in a loss of nearly all of the carbon content, 80–90% of the nitrogen content, 50% of the sulphur content, 25% of the phosphorus content and 20% of the potassium content (Raison 1979). The loss of nutrients, in the form of particulate matter and gaseous emissions, leads to the atmosphere being polluted (Lefroy et al. 1994).

Another study shows that the burning of crop stubble due to the rice–wheat cropping pattern in the northern parts of India led to the emission of 2306 Gg of carbon monoxide, 110 Gg of methane, 84 Gg of nitrous oxides and 2 Gg of nitrogen dioxide (Gupta et al. 2004).

A direct consequence of the emission of these gases is the enhanced greenhouse effect. Such a phenomenon is used to indicate an increase in temperature of the surface of the Earth, thereby causing an imbalance in the natural equilibrium. Needless to say, methane, nitrous dioxide and carbon dioxide are the major greenhouse gases, and each of these gases has a different capacity to cause an increase in the temperature of the Earth's surface, depending upon the ability of the gas to trap heat, and its average lifetime (Fairbanks et al. 2012).

Studies have also been conducted to estimate the levels of aerosol emissions due to the burning of agricultural crop residue. One such study indicates the emission of aerosols from crop residue burning is much higher than the emission from accidental forest fires. As an estimate, the burning of agricultural residue led to the emission of 663–2503 Gg per year of organic matter, 399–1529 Gg per year of organic carbon, 102–409 Gg per year of black carbon and 851–3317 Gg per year of particulate matter 2.5 (Venkatraman et al. 2006).

Another study shows that upon the burning of crop stubble, particulate material 2.5 comprised nearly 55–64 per cent of the aggregate respirable suspended particulate matter, indicating that rice burning was the major contributor and that smaller particles remained dominant. The study indicated the concentration of particulate matter 10 to be at 57 ± 15, 97 ± 21 μg per cubic metre. The study estimated the concentration of particulate matter 2.5–10 to be at 40 ± 6 μg per cubic metre. During the wheat burning season, the particulate matter concentration shot up to forty-three, fifty-one and sixty-one per cent, respectively. Further, during the rice burning season, the concentration of particulate matter increased to seventy-eight, sixty-six and seventy-one per cent, respectively (Aswathi et al. 2011).

Another study shows that the burning of agricultural crop residue also leads to the emission of polycyclic aromatic hydrocarbons. The study estimated the emission of polycyclic aromatic hydrocarbons from the burning of crop residue burning to be around 40.9 ± 15.2 µg per kilogram (Singh et al. 2013).

It can be concluded that ARCB releases various greenhouse gases and other air pollutants. GHGs trap the rays of the sun. The sun's heat is unable to escape. This greenhouse effect thereby increases the temperature of the atmosphere. The rise in Earth's temperature due to greenhouse effect leads to global warming. Crop residue burning is the main culprit of environmental pollution which affects the atmosphere, flora, fauna, wildlife and also the human health. To check and manage the same is the need of the hour.

9.4 Possible Solutions

With India's population reaching alarming numbers, the pressure on the limited non-renewable natural resources is immense. Currently, crop residue is seen as a waste since management of residue would involve collection and transportation costs. However, if utilised properly, this is a great resource which can assist in releasing the pressure on the existing limited resources. A few solutions for the management of crop residue have been discussed in this section.

9.4.1 Power/Energy Sector

The over-exploitation of resources has led to their depletion, and the economy is left with no other choice but to import natural resources like coal and petroleum. An alternative to this huge expenditure of foreign reserve could be production of energy through renewable resources. Apart from wind, solar, tidal and geothermal energy generation, production of power by utilising biomass holds great potential. Instead of merely burning the crop residue, at great detriment to the environment, the same can instead be utilised to meet the nation's growing power and energy requirements. For instance, rice straw can be used for the production of bio-ethanol, and the burning of this residue takes away the possibility of the generation of a cleaner fuel (Kumar et al. 2015). Further, unlike other renewable sources of energy, crop residue is available across India.

The Ministry of New and Renewable Energy has identified the problem created by the generation of vast amounts of crop residues and is therefore currently promoting the generation of gas from biomass such as the husk of rice, the stalks of pulses like arhar and crops like cotton and wood shavings (Hiloidharia et al. 2014). These units are envisaged to come up in rural areas so that the power generated from these gasifiers can be used to meet the requirements of industries like rice mills, where there is a need to substitute traditional fossil fuels such as diesel and coal, among others. Further, the setting up of the gasifiers in rural areas can also

9 Crop Residue Burning: Effects on Environment

help overcome the problems of insufficient electricity supply in rural India (https://mnre.gov.in/file-manager/annual-report/2017-2018/EN/pdf/chapter-3.pdf).

This programme has already been implemented in Uttar Pradesh, where the government is promoting off-grid power generation by installing three biomass gasifiers, of 1015 kW capacity in 2017–18, in an effort to meet the energy requirements of the state (https://mnre.gov.in/file-manager/annual-report/2017-2018/EN/pdf/chapter-3.pdf).

Further, another possible use of the waste generated from agriculture is to convert biomass into a cleaner cooking fuel, to be used in rural areas, where the majority of the households are still dependent on using solid fuels for the purposes of cooking. The Ministry of New and Renewable Energy has rolled out the 'Unnat Chulha Abhiyan Programme' (hereinafter, UCAP), under which biomass cooktops are being developed in an attempt to be energy efficient and cause lesser emissions as compared to conventional stoves. The Ministry has extended the UCAP to Chhattisgarh, Odisha and Mizoram in 2017–18, in an attempt to reduce the usage of solid fuels in individual households, as well as community spaces such as facilities to prepare midday meals, and government and forest rest houses (https://mnre.gov.in/file-manager/annual-report/2017-2018/EN/pdf/chapter-6.pdf).

As per the estimates by the Government of India, 100 tonnes of solid municipal waste per day has the capacity to generate 1 MW of power per day, whereas 100 tonnes of cow dung, employed daily, has the capacity to generate 1600 kg of compressed natural gas, per day. In addition to this power generation, bio-fertiliser is generated as a by-product which is extremely useful for farmers (https://mnre.gov.in/file-manager/annual-report/2017-2018/EN/pdf/chapter-7.pdf).

Statisticians expect 18,000 MW of power generation from agricultural residue and around 7,000 MW from sugar mills from sugar-producing states such as Uttar Pradesh and Maharashtra, thereby taking the expected power generation from crop residue to approximately 25,000 MW (https://mnre.gov.in/file-manager/annual-report/2016-2017/EN/pdf/3.pdf).

The potential for power generation from agricultural and agro-industrial residues is estimated at about 18,000 MW. With progressive higher steam temperature and pressure and efficient project configuration in new sugar mills and modernisation of existing ones, the potential of surplus power generation through bagasse cogeneration in sugar mills is estimated at 7,000 MW. The potential for bagasse cogeneration lies mainly in sugar-producing states, like Maharashtra and Uttar Pradesh. Thus, the total estimated biomass power potential is about 25,000 MW.

Bio-power production plants were installed in the country for the reclamation of crop residue waste into power generation.

According to a report by Ministry of New and Renewable Energy, Government of India (2016–17), an year-wise cumulative bio-power installed capacity (MW) is presented in Fig. 9.3. Data shows that up to the year 2008–09, 1751 MW bio-power capacity plants were in force, but in 2009–10 it increased to 2137 MW. There was a steady increase in the capacity generation till 2014–15 (4165 MW). In the year 2015–16, a good increase has been witnessed with the installation capacity

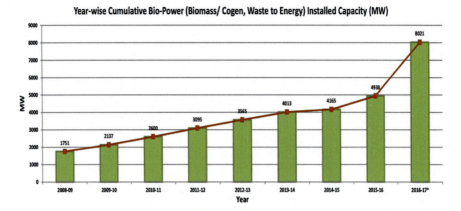

Fig. 9.3 Year-Wise Cumulative Bio-Power (Biomass/ Cogen, Waste to Energy) Installed Capacity (MW). *Source* Annual report 2016–17, Ministry of New and Renewable Energy

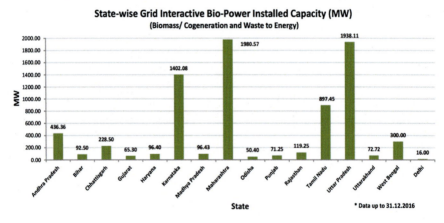

Fig. 9.4 State-Wise Grid Interactive Bio-Power Installed Capacity (MW). *Source* Annual report 2016–17, Ministry of New and Renewable Energy

increasing to 4938 MW. The situation took a better shape with a drastic increase to 8021 MW in 2016–17.

On the basis of the Annual Report (2016–17) by Ministry of New and Renewable Energy, Government of India, Fig. 9.4 illustrates the state-wise installation of bio-power production capacity from crop residue waste (biomass) in India. Maharashtra is at the top of the list with 1980.57 MW followed by Uttar Pradesh at 1938.11 MW and Karnataka at 1402.08 MW, respectively. Delhi took the last slot (16.00 MW) in this list.

However, a problem with the utilisation of the crop residue is the cost of storing and transporting the same. Farmers are not too keen on incurring the additional cost of transporting the waste to gasifiers for the generation of electricity. The author believes that farmers should be made aware of the benefits of in situ utilisation of crop residue and be encouraged to actively mix the residue with the soil while preparing the field for the next cropping season. The author also suggests that the government provides incentives to farmers who provide waste material for the production of bio-energy. These incentives could be in the form of subsidies on the purchase of agro-products, decreased interest rate on loans, etc.

9.4.2 Fodder for Animals

The high silica content of the paddy residue acts as a detriment to farmers from using the residue as fodder for their animals; however, studies indicate that if fermented the protein-rich paddy residue enhances the milk production by cattle and also increases their overall health (https://mnre.gov.in/file-manager/annual-report/2016-2017/EN/pdf/3.pdf).

9.4.3 Bedding for Animals

Using rice residue as bedding for cattle, especially in the winter months, has been tied with an improvement in the reproductive, leg and udder health of cattle and improvement in the quality of milk produced.

9.4.4 Paper Production

Agricultural residue, especially wheat and paddy straw, can be used in the paper-production industries (Jha and Sinha 2011, p. 146). Given that alarming number of trees are being cut for the production of paper, utilisation of agricultural residue would lead to a check on deforestation as well.

9.4.5 Soil Mulching

Mulch is a layer of material that is applied over the surface of the soil to conserve the soil moisture. It improves the fertility and the health of the soil and reduces the growth of weeds. Studies also show that mixing the residue of the rice crop with the soil would lead to a loamy soil which would subsequently produce a better yield of the wheat crop, by improving the pH of the soil, the water retention capacity, the density and the organic carbon content (Sidhu and Beri 2005). An in situ utilisation of the paddy husk, therefore, seems to be a viable solution.

9.4.6 Mushroom Cultivation

The straw of paddy being rich in carbon and nitrogen contents provides extremely favourable conditions for the cultivation of mushrooms (Arai et al. 2015). Mushrooms cultivated on straw accounts for one-fifth of the mushroom cultivation worldwide (Hong Van et al. 2014).

9.5 Conclusion

This chapter discusses the impact of crop residue burning on the environment. India, being an agrarian economy, is largely dependent on cultivation. However, due to the multiple cropping cycles in a year, there is hardly any time between the harvesting of one crop and the sowing of another. In order to quickly prepare the field, farmers resort to burning the crop stubble.

ARCB causes the emission of various GHGs, air pollutants and particulate matters, which adversely affects and degrades the atmosphere. Small particulate matters released due to agro-residue burning easily enter into the lungs and cause cardiac and respiratory ailments. ARCB is also responsible for decrease in soil fertility, as the water evaporates rapidly from the soil. Rise in atmospheric temperature increases the chances of pest attack on the crops.

As discussed earlier, the burning of agricultural residue leads to emission, in great quantities, of carbon dioxide, methane, sulphur and nitrous oxides. The burning also leads to emission of aerosols. Rice stubble is the major culprit and contributes to the maximum emission of gaseous and particulate matter. The imbalance leading to increasing temperatures is a matter of grave concern.

The need of the hour is to reduce the emissions of GHGs and also to manage and reclaim the surplus crop residue properly. The stepping stone for the same is to convince the farmers to stop burning crop stubble in their fields. To overcome this big threat of our planet, surplus crop residue must be managed properly. The reclamation should be suitable to improve the economic condition of the country and reduce atmospheric pollution, while being cost effective.

The author has explored how crop residue has a massive potential to be used as a resource. Crop residue can be used for electricity/power generation, as a natural fuel, as fodder and bedding for animals, in mushroom cultivation, among others. On- and off-site utilisation of this so-called waste material would not only lead to a decrease in the detriment to the environment, but would also lead to an increase in the limited resources on the planet.

References

Arai H, Hosen Y, Hong VNP, Thi NT, Huu CN, Inubushi K (2015) Greenhouse gas emissions from rice straw burning and straw-mushroom cultivation in a triple rice cropping system in the Mekong Delta. Soil Sci Plant Nutr 61(4):719–735. https://doi.org/10.1080/00380768.2015.1041862

Aswathi A et al (2011) Study of size and mass distribution of particulate matter due to crop residue burning with seasonal variation in rural areas of Punjab, India. J Environ Monit 13(4):1073–1081

Badve VC (1991) Feeding systems and problems in the Indo-Ganges plain: case study. In: Speedy A, Sansoucy R (eds) Feeding dairy cows in the tropics. Proceedings of the FAO expert consultation held in Bangkok, Thailand, July 7–11, 1989

Baede APM, Ahlonsou E, Ding Y, Schimel D The climate system: an Overview, IPCC Reports, p 3. https://www.ipcc.ch/ipccreports/tar/wg1/pdf/TAR-01.PDF

Damodaran H (2017) Delhi air pollution: a (crop) burning issue, and the way out, The Indian Express, 10 Nov 2017. http://indianexpress.com/article/explained/delhi-air-pollution-smog-crop-stubble-burning-delhi-school-shut-odd-even-scheme-manish-sisodia-amarinder-singh-arvind-kejriwal-punjab-farmers-4930457/

Deshpande T (2017) State of agriculture in India, PRS Legislative Research, Mar 2017. http://www.prsindia.org/administrator/uploads/general/1517552563 ∼ ∼ State%20of%20Agriculture%20in%20India.pdf

Devi S, Gupta C, Jat SL, Parmar MS (2017) Crop residue recycling for economic and environmental sustainability: the case of India. Open Agric, De Gruyter Open 2:486–494. https://www.degruyter.com/view/j/opag.2017.2.issue-1/opag-2017-0053/opag-2017-0053.xml

Economic Survey 2017–18, Volume I, Government of India ministry of finance department of economic affairs economic division Jan 2018, p 101. http://www.indiaenvironmentportal.org.in/files/file/economic%20survey%202017-18%20-%20vol.1.pdf

Fairbanks M, Bowran D, Pasqual G (2012) Agricultural greenhouse gas emissions. Department of Agriculture and Food, Western Australia, Perth. Bulletin 4837

Gadde B, Bonnet S, Menke C, Garivait S (2009) Air pollutant emissions from rice straw open field burning in India, Thailand and the Philippines. Environ Pollut 157:1554–1558

Government of India, Annual Report 2016–17, Ministry of New and Renewable Energy, Ch 3, P. 14 https://mnre.gov.in/file-manager/annual-report/2016-2017/EN/pdf/3.pdf

Government of India, Annual report 2016–17, ministry of new and renewable energy, P 36. https://mnre.gov.in/file-manager/annual-report/2016-2017/EN/pdf/3.pdf

Government of India, Annual report 2017–18, ministry of new and renewable energy, Ch 3, P 28. https://mnre.gov.in/file-manager/annual-report/2017-2018/EN/pdf/chapter-3.pdf

Government of India, Annual report 2017–18, ministry of new and renewable energy, Ch 6, P. 80. https://mnre.gov.in/file-manager/annual-report/2017-2018/EN/pdf/chapter-6.pdf

Gupta PK et al (2004) Residue burning in rice-wheat cropping system: cause and implications. Curr Sci 87(12):1713–1717

Gupta PK, Sahai S, Singh N, Dixit CK, Singh DP, Sharma C (2004) Residue burning in rice-wheat cropping system: causes and implications. Curr Sci 87(12):1713–1715

Hiloidharia M, Das D, Baruah DC (2014) Bioenergy potential from crop residue biomass in India. Renew Sustain Energy Rev 32:504–512, P 505. https://www.sciencedirect.com/science/article/pii/S1364032114000367

Hobbs P, Morris M (1996) Meeting South Asia's future food requirements from rice-wheat cropping systems: priority issues facing researchers in the post-green revolution era', NRG paper 96(01)

Hong Van NP, Nga TT, Arai H, Hosen Y, Chiem NH, Inubushi K (2014) Rice straw management by farmers in a triple rice production system in the Mekong Delta, Vietnam. Trop Agr Develop 58:155–162

Id at Ch 7, P 83. https://mnre.gov.in/file-manager/annual-report/2017-2018/EN/pdf/chapter-7.pdf

Jha P, Sinha A (2011) Application of rice-straw as raw material for production of handmade paper. IPPTA: Q J Indian Pulp Paper Tech Assoc 23(2):145–148

Karlen DL et al (1994) Crop residue effects on soil quality following 10-years of no-till corn. Soil Tillage Res 31(2–3):149–167

Kumar P et al (2015) Socioeconomic and environmental implications of agricultural residue burning, Springer Briefs in Environmental Science, p 13. https://doi.org/10.1007/978-81-322-2014-5_2

Kumar P, Kumar S, Joshi L (2015) Socioeconomic and environmental implications of agricultural residue burning a case study of Punjab, India. Springer, New Delhi Heidelberg, New York Dordrecht London, P 72

Lefroy RD, Chaitep W, Blair GJ (1994) Release of sulphur from rice residue under flooded and nonflooded soil conditions. Aust J Agric Res 45:657–667

Meshram JR (2002) Biomass resources assessment programme and prospects of biomass as an energy resource in India. IREDA News 13(4):21–29

NASA Visible Earth, Stubble Burning in India, 30 Oct 2014. https://visibleearth.nasa.gov/view.php?id=84680

Raison RJ (1979) Modification of the soil environment by vegetation fires, with particular reference to nitrogen transformation: a review. Plant Soil 51:73–108

Sidhu BS, Beri V (2005) Experience with managing rice residues in intensive rice-wheat cropping system in Punjab. In: Abrol IP, Gupta RK, Malik RK (eds) Conservation agriculture: status and prospects. Centre for Advancement of Sustainable Agriculture, National Agriculture Science Centre, New Delhi, pp 55–63

Singh DP et al (2013) Emissions estimates of PAH from biomass fuels used in rural sector of Indo-Gangetic plains of India. Atmos Environ 68:120–126

Singh RP, Dhaliwal HS, Sidhu HS, Manpreet-Singh YS, Blackwell J (2008) Economic assessment of the happy seeder for rice-wheat systems in Punjab, India. In: Conference paper, AARES 52nd annual conference. ACT, Canberra, Australia

Technology Information Forecasting and Assessment Council (1991) Techno market survey on "Utilization of agriculture residue (farms and processes)". Department of Science and Technology, New Delhi

Venkatraman C et al (2006) Emissions from open biomass burning in India: integrating the inventory approach with higher solution moderate resolution imaging spectroradiometer (MODIS) active fire and land count data. Glob Biogeochem Cycles 20(GB2013-20)

Yevich R, Logan JA (2003) An assessment of biofuels use and burning of agricultural waste in the developing world. Glob Biogeochem Cycles 17. https://doi.org/10.1029/2002GB001952

Chapter 10
Rooftop Solar Power Generation: An Opportunity to Reduce Greenhouse Gas Emissions

Neeru Bansal, V. K. Srivastava and Juzer Kheraluwala

Abstract Solar energy is being promoted in India as one of the main components of renewable energy as the country receives good solar radiation over 300 days a year. It has emerged as a potential green alternative to address climate change issues by reducing reliance on conventional fossil fuel-based energy. The government has taken many policy initiatives to promote solar power generation and aims to produce 100 GW of solar power by the year 2022, out of which 40 GW is planned from solar rooftops. The land requirement for solar power generation systems is large, and in urban areas, acts as a major constraint. Rooftop solar power generation systems are an option and opportunity under such circumstances. This chapter focusses on the opportunities available to adopt rooftop solar power generation in the residential sector. The constraints in adopting these systems and the factors influencing decision of the household for installation of such systems are discussed. The primary data for the study has been collected through interviews with the households and discussions with the stakeholders—policy makers, distribution companies, consultants and the households with access and ownership rights to rooftops. The secondary data analysed includes published as well as unpublished data. The analysis highlights that there are numerous opportunities available to adopt rooftop solar power generation systems. These are mainly in the form of favourable policy instruments leading to economic gains. The reduction in the energy bill is the most important factor influencing decision of the household to adopt rooftop solar power generation. The system becomes attractive if government subsidies are available. The positive recommendation from friends and family who have already installed these systems drive the final decision of the household. The

---

N. Bansal (✉)
CEPT University, Ahmedabad, Gujarat, India
e-mail: bansal.neeru@cept.ac.in

V. K. Srivastava
Sankalchand Patel University, Visnagar, Gujarat, India
e-mail: drvks9@gmail.com

J. Kheraluwala
Ernst & Young LLP, New Delhi, India
e-mail: juzer50@gmail.com

© Springer Nature Singapore Pte Ltd. 2019
N. Shurpali et al. (eds.), *Greenhouse Gas Emissions*, Energy, Environment,
and Sustainability, https://doi.org/10.1007/978-981-13-3272-2_10

reduction in greenhouse gas emissions and carbon footprint from adoption of rooftop solar power generation systems have been discussed in the chapter. The findings lead to policy recommendations to promote rooftop solar power generation and making it more attractive.

Keywords Rooftop solar power generation · Greenhouse gas reductions Opportunities · Constraints

10.1 Introduction

Solar energy is being promoted in India as one of the main components of renewable energy. The country receives good solar radiation of 4–7 kWh m^{-2} day^{-1} for over 300 days a year. Solar energy has emerged as a potential green alternative to address emission of greenhouse gases (GHGs) and the resultant climate change issues by reducing reliance on conventional fossil fuel-based energy. The government has taken many policy initiatives to promote solar power generation and aims to produce 100 GW of solar power by the year 2022, out of which 40 GW is planned from solar rooftops. The land requirement for solar power generation systems is high and acts as a major constraint in urban areas. The rooftop solar power generation systems are an alternative and an opportunity for generating power right at the consumer end. The rooftop solar power generation has been focused upon by many countries like Germany and Japan, and special policy initiatives have been rolled out to promote this sector.

The growth of rooftop solar power generation systems is directly linked to reduction in GHGs at the point of consumption itself. In India, the solar power generation is witnessing a good growth rate as the share of solar power used to be only 0.02% (0.03 GW) in 2009 (prior to the launch of Jawaharlal Nehru National Solar Mission (JNNSM) in 2011) and has increased to 5% in 2018. The share of rooftop solar power generation is still in a nascent stage at 0.25%. The government has taken many initiatives in terms of policy instruments, fiscal incentives and market innovations to promote this sector. This chapter focusses on analysing the opportunities available to adopt rooftop solar power generation, and residential sector is the focus, reason being that the households have the power to take the final decision in favour of or against the installation of rooftop solar power generation system.

10.2 Data Collection and Methodology

India has got a federal structure, and the central government has offered several incentives to promote and encourage the growth of solar power in the country; coupled with this are the incentives offered by the respective state governments.

Based on the solar irradiance, the states of Gujarat, Tamil Nadu and Rajasthan have capitalized on solar energy reserve by taking policy initiatives at the state level for promoting renewable energy. Gujarat has been the first state in India to develop a solar policy in 2009, two years before the commencement of JNNSM. This provided Gujarat a head start in the generation of renewable energy. In Gujarat, as per the data of Gujarat Energy Development Agency (GEDA),[1] Ahmedabad has led the way for commissioning of rooftop solar power projects followed by Vadodara, Surat and Rajkot. For this study, Ahmedabad has been considered as the study area. The statistics for rooftop solar power generation projects in the state have shown that the maximum projects are in the residential sector, followed by the government buildings, industrial buildings and commercial buildings (Ajay 2018). This leads to selection of residential sector as focus area for this study.

10.2.1 City Profile—Ahmedabad

Ahmedabad is the largest city in the state of Gujarat. The estimated population of the city as per the latest census records of 2011 is 5.63 million, and it is the seventh most populous city in the country (Government of India 2011). The city covers an area of 464.17 km^2, and the Sabarmati River divides it into eastern and western Ahmedabad. The city has a very diverse housing typology. The old city has historic Pol[2] culture and is on eastern side. The typology of residential buildings has diversified with time into individual bungalows, row houses, tenements and multi-storey apartments (Fig. 10.1).

The city is located at a latitude and longitude of 23.03°N and 72.58°E and receives an average daily horizontal global irradiation (GHI) of 5.33 kWh m^{-2}, which is relatively high when compared to other major Indian cities. The monthly variation in GHI for Ahmedabad is shown in Table 10.1. The months of April and May receive a GHI greater than 7 kW m^{-2} day^{-1}. The lower GHI in the months of July and August is due to the onset of the monsoon.

The city receives uninterrupted power supply throughout the year with power outages and voltage fluctuations almost unheard of. This may have a direct correlation on the need felt for installing rooftop solar by residential consumers in the city. There are two power distribution companies (DISCOMs) supplying power to the city—Torrent Power Ltd., which supplies power to the areas under the jurisdiction of Ahmedabad Municipal Corporation (AMC), and Uttar Gujarat Vij Company Ltd. (UGVCL), which supplies power to the peripheral parts of the city under the jurisdiction of Ahmedabad Urban Development Authority (AUDA). The total power supply in Ahmedabad has risen at a compound annual growth rate

---

[1]It is the nodal agency in Gujarat for the renewal energy development in the state.

[2]A pol is a housing cluster which comprises many families of a particular group linked by caste and religion.

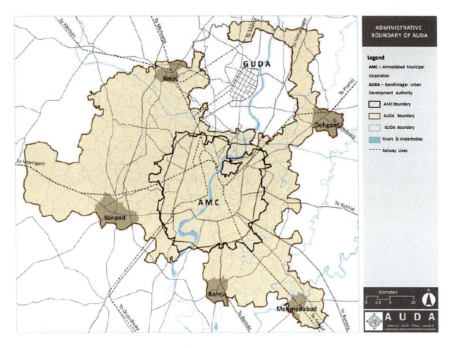

Fig. 10.1 Administrative jurisdiction in Ahmedabad

Table 10.1 Monthly average global horizontal irradiance (GHI) for Ahmedabad

Month	Jan	Feb	Mar	Apr	May	Jun	Jul	Aug	Sep	Oct	Nov	Dec
GHI kWh/ m^2/day	4.81	5.76	6.63	7.28	7.54	6.5	4.98	4.63	5.69	5.87	5.02	4.5

Source (Ezysolare 2016)

(CAGR) of 4.5% per annum from 2010 to 2018 (Torrent Power Limited 2017). It can be observed (Fig. 10.2) that there is a difference between the total power supplied and the total power billed. The difference represents transmission losses between the point of generation and the point of consumption. The average transmission losses since 2010 have hovered around 9%, which is much lower than the national average of 26% (The Times of India 2018).

10.2.2 Data Collection

The data collection for this study was carried out between November 2017 and April 2018 through primary and secondary surveys. The secondary data has been

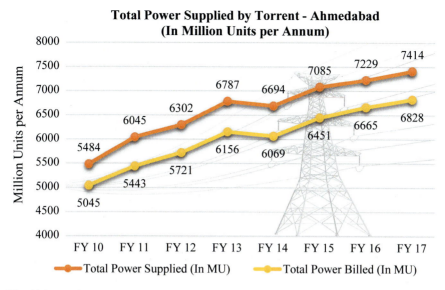

Fig. 10.2 Total power supplied by torrent in Ahmedabad

collected through published reports, journal articles and news items. In addition, unpublished data from GEDA for rooftop power systems has been collected and analysed. Interviews have been conducted with the stakeholders—regulatory agency (GEDA), power distribution company (Torrent Power Ltd.) and Channel Partners.[3]

The interaction with GEDA revealed that 577 households have installed rooftop solar PV systems in Ahmedabad till 20 December 2018. These have been installed under the 'solar rooftop project' of the state government, which was started in November 2016 under the Gujarat Solar Policy, 2015.

The split-up between the number of consumers connected to Torrent Power Ltd. and UGVCL is 478 and 99 consumers, respectively. A further analysis of the data provided by GEDA shows that more than 80% of the installed rooftop solar systems are on the western side of Ahmedabad; certain areas have higher concentration of these systems, e.g. Sola and Science City area, Ambawadi, Paldi, Satellite, Thaltej, Bodakdev and Vastrapur. These areas have been primarily focused during the process of sampling for household surveys. The interactions with the channel partners have informed that approx. 97% of the rooftop solar photovoltaic installations in the residential sector are in individual houses/bungalows/villas. The major reason for the poor response from consumers in apartments is due to the

---

[3]These are the authorized consultants for commissioning of rooftop solar power generation systems. They are authorized to carry out all the activities—from estimating the potential of solar power generation to making an application for subsidies, setting up the rooftop solar PV system and maintaining it for five years.

Fig. 10.3 View of neighbourhood from rooftop of high-rise building

eligibility criteria mandated by GEDA for giving subsidies for installation of rooftop solar power generation systems in residential sector. This criterion mandates legal possession of the rooftop; however, in apartments, the residents have a joint ownership of the rooftop, and this necessitates a No Objection Certificate (NOC) from all the flat owners which is difficult to get. This has been cited as the biggest constraint in adoption of rooftop solar PV systems among residential apartments and towers.

The primary data has been collected between November 2017 and April 2018 through household surveys. The household surveys consisted of a questionnaire, filled by the surveyor through an open-ended, semi-structured interview. Initially, the households were approached through contacts given by the channel partners. The sample size was further expanded through the chain referral technique (Explorable Homepage 2018). This technique is used by researchers to identify potential candidates in a study where it is difficult to find or locate candidates. This technique has been very successful as it helped in obtaining approx. 60% of the samples. During the later phase of the primary surveys, households with solar systems have also been identified from terraces of high-rise buildings. The terraces provided an excellent, unobstructed view of the entire neighbourhood (Fig. 10.3).

The interviews were conducted at the homes of the respondents after taking prior appointments. The samples have been divided into two cohorts—households with (25 samples) and households without rooftop solar (30 samples). The methodology for household surveys is represented in Fig. 10.4. The building typology of the surveyed households is represented in Fig. 10.5.

The data collected has been analysed to find out the opportunities available for installation of rooftop solar power generation systems. The data has also been analysed to find out the constraints, if any, for the adoption of these systems.

10 Rooftop Solar Power Generation: An Opportunity to Reduce … 147

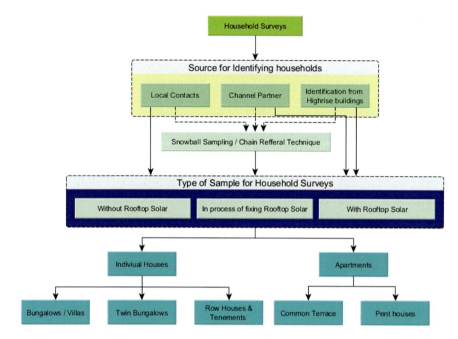

Fig. 10.4 Household survey methodology

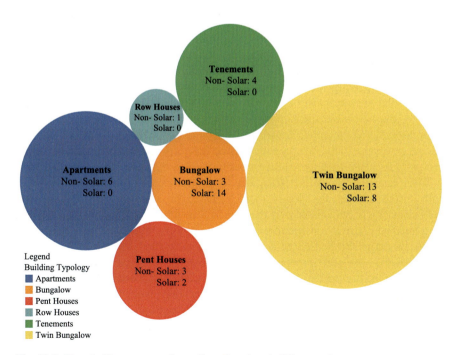

Fig. 10.5 Household survey samples: split-up based on building typology

10.3 Opportunities for Rooftop Solar Power Generation

To understand the opportunities available to adopt rooftop solar power generation systems, the primary data and secondary data have been analysed. The findings from the data analysis are categorized into following thematic areas:

- Enabling policy instruments
- Mandatory project clearances
- Reduction in carbon footprint
- Economic gains.

10.3.1 Enabling Solar Policy Instruments

In order to promote rooftop solar power generation systems[4] and reduce the emission of GHGs, and to honour the country's commitment to global initiatives, the central and the state governments have devised certain enabling policies. Some of these policy initiatives focus especially on the consumer end. These policy instruments focus on incentivizing the end-users to adopt rooftop solar power generation systems and have been discussed in the following paragraphs.

Net-metering. Net-metering is the concept where the consumers are connected to the DISCOM's grid and provided with a bidirectional meter, which tracks the net power consumed by the consumer. The power generated from the rooftop solar power generation system is first consumed within the household, and any unused surplus electricity is fed into the DISCOM's grid (Tongsopit et al. 2016; Shivalkar et al. 2015). When the consumer's load is higher than the power produced by the rooftop solar system, power is consumed from the DISCOM. The consumer only pays for the net power consumed, which is difference between power consumed from the grid and the power fed into the grid. For example, if the rooftop solar PV system generates 1000 kWh per billing cycle and the total power imported from the grid is 1250 kWh, then the consumer will be billed only for 250 kWh.

The Gujarat Solar Policy relies heavily on net-metering for small consumers with installed capacities lesser than 1 MW. The solar rooftop project with net-metering was implemented by the state in November 2016. It initially received a very tepid response as most residential consumers were following a wait-and-watch approach. From the household surveys, we observe that majority of the rooftop solar systems have been installed after November 2017. This can be attributed to an amendment in the policy. Earlier there used to be a cap on the capacity of the rooftop solar power generation at 50% of the sanctioned load. This was revised, and the residential consumers were allowed to install rooftop solar PV systems up to

---

[4]All the rooftop solar power generation systems in the study area have photovoltaic (PV) systems.

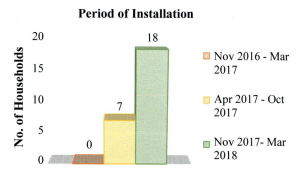

Fig. 10.6 Period of rooftop solar PV system installation

100% of their sanctioned load. The removal of this cap has enhanced investor sentiment by making rooftop solar an attractive investment option (Fig. 10.6).

Net-metering for rooftop solar in Gujarat has been coupled with another policy instrument popularly known as feed-in tariff (FIT). This is an added incentive to the consumers who now have the option of getting paid for any surplus power which they export to the grid, at the end of bimonthly billing cycle. The current rate for surplus power fed into the grid is Rs. 3.22/kWh.

Another key parameter of Gujarat's policy on net-metering is that there is no cap on the quantum of power being fed into the grid when compared to power consumed from the gird. Most states specify that the total power supplied into the grid must not exceed 90% of total power consumed from the grid over a period of 12 months. Excess power generated in certain months is carried forwarded and settled annually. There are a few states where excess power generated beyond the prescribed limit of 90% turns void at the end of the settlement period. However, in Gujarat, the consumers are paid for every excess unit fed into the grid and the settlement is done on a bimonthly basis rather than an annual basis. This is a great incentive and an economic opportunity.

Subsidies for Installation of Rooftop Solar Power Systems. A crucial demand-side policy instrument is the provision of subsidies for the installation of rooftop solar power generation systems. These subsidies have significantly reduced the total installation costs and are available only to residential consumers. The consumers in the commercial and industrial sectors are provided with fiscal incentives in the forms of tax benefits and accelerated depreciation. The GEDA has fixed the cost of 1 kW solar PV system at Rs. 69,000. This cost includes all equipment costs, installation costs and the operation and maintenance (O&M for 5 years) costs. There are two different capital subsidies, one by the Ministry of New and Renewable Energy (MNRE) and another by the Government of Gujarat (GOG) to the residential sector. The MNRE offers a direct 30% subsidy on the total installation cost per kW system up to a limit of 50 kW system. The GOG offers a subsidy of Rs. 10,000 per kW up to a maximum of 2 kW system. This means that the residential consumer/beneficiary is required to pay only Rs. 38,300 per kW. These capital subsidies have enabled rooftop solar systems to be more attractive financially to residential consumers.

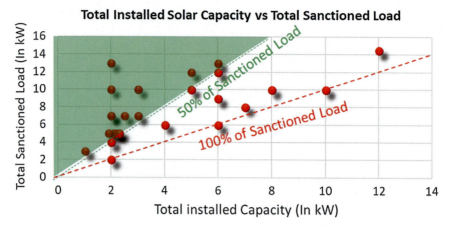

Fig. 10.7 Total installed rooftop solar capacity versus sanctioned load

The impact of these subsidies has been huge. The household surveys have revealed that most of the consumers have installed 2 kW rooftop solar PV systems due to the cap on the state subsidy as shown (Fig. 10.7), though they have roof areas and sanctioned loads which can support larger capacity systems. There are only three households which have installed solar power system having capacity equal to the sanctioned load. This also means that the rooftop solar systems are being found attractive by the households only if the subsidies are available. Hagerman et al. (2015) has observed that even in the USA, for the residential consumers, the solar PV is not able to achieve socket parity without subsidies in most of the states. At the same time, through the policy instrument of 'subsidy', the government in a way is delegating its obligation of reducing emission of GHGs, to the consumers. The consumer is now a direct contributor towards reducing the carbon footprint of the country.

Inclusion of the Cost of all Components in Capital Cost of the System. The policy has another enabling instrument, i.e. inclusion of the cost of all components (solar panels, inverter, wiring and cabling, cost of bidirectional meter, mounting structure, etc.) in the capital cost of solar PV system, fixed by GEDA at Rs. 69,000 per kW. The cost is also inclusive of government taxes and the profit margin of the channel partners who install the system. This has ensured that all the costs associated with installation of rooftop solar PV systems remain transparent and easy to understand. The only additional cost to be borne by the consumer is the connectivity charges payable to their respective DISCOM. This charge is variable based on the consumer type and the DISCOM, and hence is not included in the overall cost of the system.

By ensuring that all costs for the system are included, the government has ensured that there are no price negotiations. Any unscrupulous activity on part of the channel partners like demanding additional costs for components can be easily avoided. By providing a transparent system of capital costs break-up, the confusion and doubts arising in the minds of the consumers, which generally leads to avoidance and reluctance to adopt to new technologies, have been converted into an opportunity as the policy has ensured smooth and simple transactions.

Inclusion of Operation and Maintenance Costs. The Gujarat Solar Policy has provided an additional benefit to the consumers which includes free operation and maintenance (O&M) of the installed rooftop solar PV system for a period of five years. This is another important demand-side enabling policy instrument. As per the policy, the channel partner is responsible for the O&M of the system which also includes the replacement of PV and BOS[5] components due to any manufacturing defects. This policy instrument has further enhanced consumer confidence as the systems pays for itself with the free O&M for a period of five years. The reluctance of the consumers with respect to maintenance issues of the rooftop solar system before it achieves its payback period has been smartly negated by this policy instrument. Further, the importance of O&M will be well understood by the consumers in the first five years, and they will be willing to avail O&M contracts for the remaining period of 20 years. This is a win-win situation for the channel partners, as they will be ensured of O&M-related business for the entire period of Power Purchase Agreement (PPA).

The mix of the policy instruments discussed in above paragraphs is a great opportunity and a game-changer in encouraging installation of rooftop solar PV systems across residential consumers.

10.3.2 Rooftop Solar Power Generation and the Project Clearances

Building Plan Clearance. The Government of India has formulated guidelines such as (i) Model Building Bye-Law, 2016, prepared by the Town and Country Planning Organization and (ii) Energy Conservation Building Code (ECBC), 2017, prepared by the Bureau of Energy Efficiency under the Ministry of Power. Both these guidelines promote and mandate the installation of rooftop solar PV systems. The guidelines provide immense opportunities for development of rooftop solar in urban areas.

The Model Building Bye-Laws (MBBL) have focused on the inclusion of solar power generation on rooftops of buildings. The guidelines for the installation of rooftop solar are applicable to certain buildings based on their size and their power consumption. The section of the bye-laws with respect to rooftop solar has been adopted by most Indian states including Gujarat. The relevant sections for installation of rooftop solar PV systems have been reproduced in Table 10.2. These norms are applicable for both new and existing buildings, though implementation in existing buildings may not be very practical.

---

[5]BOS—All components of a solar PV system apart from the photovoltaic modules is known as Balance of Systems.

Table 10.2 Guidelines for rooftop solar power installation (Town and Country Planning Organization 2016)

S. no	Category of building	Area standards	Generation requirement[a]
Residential			
1	Plotted housing	For HIG plots and above	Minimum 5% of connected load or 20 W sq. ft^{-1} for available roof space[b], whichever is lesser
2	Group housing	All proposals, as per group housing norms	Minimum 5% of connected load or 20 W sq. ft^{-1} for available roof space[b], whichever is lesser
All other buildings having shadow-free rooftop area >50 m^2			
3	Educational	Plot size 500 m^2 and above	Minimum 5% of connected load or 20 W sq. ft^{-1} for available roof space[b], whichever is lesser
4	Institutional		
5	Commercial		
6	Industrial		
7	Mercantile		
8	Recreational		

[a]Area provisions on rooftop shall be 12 m^2 per 1 kW, as suggested by MNRE
[b]'Available roof area' = 70% of total roof size, considering 30% area reserved for residential amenities

The norms provided in the bye-laws translate to a very minimal requirement of rooftop solar to be installed on the building premises and may not result into any substantial reduction in building energy consumption or reduction in carbon footprints. However, these norms are likely to create a conducive environment for improving awareness among households for the requirement of rooftop solar power generation. The formulation of these norms on a national level will set the precedent among states governments and urban local bodies to modify and update their rooftop solar guidelines, thereby creating additional opportunities in the rooftop solar sector in the urban areas. Once the strict monitoring of these norms is implemented during process of building plan approval, higher adaptation of rooftop solar power generation system can be achieved.

Environmental Clearance. To incorporate necessary environmental safeguards at project planning stage and not as an afterthought, certain projects require environmental clearance (EC). These projects include 'building and construction projects' and 'townships and area development projects'. The projects are required to conduct environment impact assessment (EIA) to identify potential environmental impacts and propose mitigation strategies along with traditional technical and financial evaluations. The scale of the project and the associated nodal agency, which will issue environmental clearance, is described in Table 10.3.

The law requires 'Environmental Cell' to be constituted at the local authority level and integrate environmental conditions with the building permission being granted by the local authorities. The local authorities are required to certify the compliance of the environmental conditions prior to issuance of completion certificate based on the recommendation of the Environmental Cell. The environmental

10 Rooftop Solar Power Generation: An Opportunity to Reduce ...

Table 10.3 Environmental clearances required for building approval

S. no	Type of project	Scale of project	Nodal agency	Remarks
1	Building and construction projects	$\geq 20{,}000$ and $<1{,}50{,}000$ m^2 of built-up area	Local body	The term built-up area for the purpose of this notification is the built-up or covered area on all floors put together including its basement and other service areas, which are proposed in the buildings and construction projects
2	Townships and area development projects	$\geq 1{,}50{,}000$ and $<3{,}00{,}000$ m^2 built-up area	State level (SEIAA)	
3	Townships and area development projects	$\geq 3{,}00{,}000$ m^2 built-up area Or covering an area ≥ 150 ha	Central government (MOEF)	

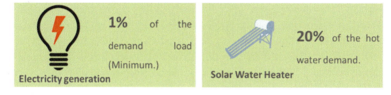

Fig. 10.8 Mandatory limits enforced by EC for renewable energy in buildings

conditions to be integrated in building bye-laws for renewable energy are shown in Fig. 10.8.

The mandatory percentage for renewable energy in this case is shown in Fig. 10.8, though a small percentage can act as an opportunity to set up solar rooftop power systems. Once enforced, there is every possibility that the roof owners would like to exploit the full potential of power generation and may not limit it to the mandatory limit.

10.3.3 Reduction in Carbon Footprint

The solar power generation systems are an environmentally friendly option for enabling the reduction of GHGs which otherwise would have been emitted due to power produced from conventional fossil fuel-based power plants. The power consumers now have the opportunity to directly contribute to a reduction in their carbon footprints through installation of rooftop solar PV systems. The household surveys and the analysis have revealed that the installation of rooftop solar PV systems results in reduction of CO_2 emission by 1.46 MT of CO_2 per person per

Table 10.4 Renewable purchase obligations

Year	Minimum quantum of purchase (%) from renewable energy sources			
	Wind (%)	Solar (%)	Others—biomass, bagasse, hydro and MSW (%)	Total (%)
2017–18	7.85	3	0.5	11.35
2018–19	7.95	4.25	0.5	12.7
2019–20	8.05	5.5	0.75	14.3
2020–21	8.15	6.75	0.75	15.65
2021–22	8.25	8	0.75	17

Source Gujarat Electricity Regulatory Commission (GERC) (2017)

annum in Ahmedabad. This is roughly equivalent to what a single tree absorbs over its entire lifetime.

The power generation from rooftop solar PV is currently only 0.25% of the total power which means that there is massive potential in the future for reduction of GHGs through this mechanism. Moreover, DISCOMs across the country have now been given renewable purchase obligation (RPO) targets. This step will ensure that certain percentage of their power is generated from renewable energy sources. The RPO for DISCOMs and captive power producers in Gujarat for financial year 2018 mandates that 12.7% of total power must be from renewable energy sources, which includes 4.25% of the power generated from solar energy. This provides a great opportunity to promote rooftop solar power generation through RPO of the DISCOMs as shown in Table 10.4.

10.3.4 Economic Gains from the Use of Rooftops for Solar Power Generation

The rooftops of residential buildings in urban areas lie relatively unused for most parts of the year. The household surveys have revealed that 57% of the residential rooftops are not used for any purpose. A further 30% of the respondents use the rooftop for purposes which are of no economic value, such as drying clothes and kite flying (which happens only for a day or two in a year). A small number of rooftops have installed solar water heater systems, which replace electricity and gas as an alternate heating method. An assessment has been made in this study for the potential economic gains if rooftop power generation systems are installed by the households (Fig. 10. 9).

There are two important factors which determine the capacity of rooftop solar power generation system to be installed—(1) sanctioned load and (2) shadow-free area. The sanctioned load is important as one can install the maximum capacity of

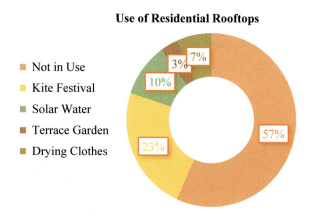

Fig. 10.9 Rooftop use of respondents from household surveys

the system to be equal to sanctioned load, even if excess roof area is available. It is important to calculate the 'shadow-free area' on the roof for the reason that the solar power panels are arranged in 'series'. Any shadow on the panel breaks the circuit, and therefore, no power is generated.

As a case study, a residential household has been selected as the pilot project to assess the potential of solar power generation. The details are as below:

- Latitude: 23.0345°N
- Longitude: 72.5529°E
- Area: Ahmedabad
- Total Rooftop Area: 1,872 sq. ft
- Total Sanctioned Load: 11.7 kW
- Electricity Consumption: 1642 units (November and December).

In this case, if the available roof area is considered, then the solar power generation of 15 kW can be installed. This potential gets reduced to 11.7 kW, due to the limit imposed by the policy that the maximum installed capacity shall not exceed 100% of the sanctioned load. The shadow analysis for the selected case has shown that the shadow-free area available is only 750 sq. ft, which can accommodate solar power generation system of 6 kW. Therefore, though the available roof space can accommodate a system of 15 kW but in actual, only a system of 6 kW can be installed. This system of 6 kW will generate 1312 units, based on solar irradiance, for the months of November and December. The residential unit will be able to compensate 75% of the power consumed by it from rooftop solar power generation system. This will result into an economic gain of Rs. 8922 for the two months under consideration. If we consider the existing cost of I kW at Rs. 69,000 and build in the current subsidies extended by the central and the state governments, the payback period for the system is just 5 years and 9 months, whereas the total life of the system is 25 years. Similar observations have been made by (Dube et al. 2017) in their study focused on performance assessment of 1 KW solar rooftop systems. Therefore, the assessment of the rooftop solar power generation system

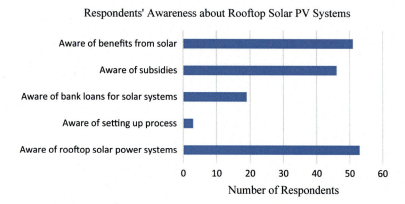

Fig. 10.10 Awareness level of respondents about rooftop solar systems

indicates that the economic potential of an unused roof is very high. Adding on to this is the positive impact on the environment in terms of reduction in GHGs (Shukla et al. 2017).

10.4 Constraints in Adopting Rooftop Solar Systems

10.4.1 Consumer Awareness

As discussed in the previous section, there are many opportunities available with the residential sector to install rooftop solar power generation systems. However, not many households have adopted it. An attempt has been made in this section to understand the awareness level of the consumers with respect to rooftop solar power generation systems, which might be acting as a major constraint in adoption of rooftop solar power generation systems. The respondents during the household surveys have been questioned about the awareness levels with respect to the following—awareness about rooftop solar PV systems, process of setting up rooftop solar system, availability of the bank loans for solar systems, availability of the subsidies by the government and awareness about economic and environmental gains from the rooftop solar power generation. The responses have been graphically represented in Fig. 10.11 (Fig. 10.10).

The analysis shows that 95% of the respondents are aware of the rooftop solar power generation. They are also aware that these systems could lead to certain economic and environmental benefits, but the quantum of these gains is not known. Kappagantua et al. (2015) have observed that during personal interviews, the respondents generally pay attention to details on net and gross metering as these have a direct impact on the economic gains and the payback periods. The respondents are also aware that government provides certain subsidies. However, most of the

Fig. 10.11 Mode by which consumers became aware about rooftop solar systems

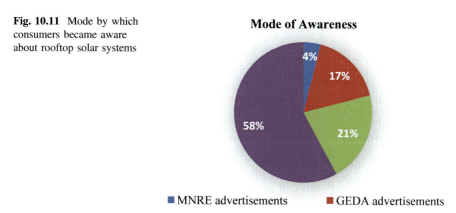

respondents are not aware of the setting up process for rooftop solar power generation system. The respondents are also not aware that the cost of the solar power generation systems can be included in the home improvement loans.

Most of the respondents came to know about the rooftop solar power systems through word of mouth, from friends and family. A very small number of the respondents came to know about installation of solar PV systems through awareness campaigns and advertisements by MNRE and GEDA. This means that people rely on the experience and recommendation of their acquaintances to take a final decision in favour of or against the installation of these systems. The households still lack the confidence for investing in solar rooftop systems, based only on promotion campaigns of the government. They would rather wait until someone in the vicinity or someone they know has set up the system and has tested its outcome. The interactions revealed that most of the consumers have installed rooftop solar PV systems primarily for the purpose of reducing their electricity bills.

10.4.2 Rooftop Becoming Unusable

Another constraint which has deterred consumers from installing rooftop solar systems is that their terrace would be rendered unusable for other purposes once the system is installed. A sizeable number of respondents have cited that their rooftop area would become completely unusable after the installation of a rooftop solar PV system. The study (Thomas 2018) looked into households which have installed rooftop solar PV systems, and it has been assessed that almost 400–900 sq. ft of roof area is still freely available after the installation of the system. This perceived constraint is common among respondents who have not installed a rooftop solar PV system. Figure 10.12 shows the available rooftop area of each individual household after they have installed a rooftop solar PV system.

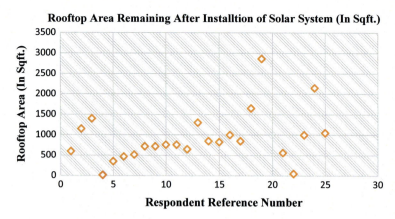

Fig. 10.12 Rooftop area remaining after installation of rooftop solar PV system

10.4.3 Reliable Power Supply

The Torrent Power Limited provides a reliable power supply in Ahmedabad. There are almost no power outages or voltage fluctuations observed in the city. This is a very positive scenario. However, from the perspective of adopting rooftop solar power generation system, this sometimes acts as a constraint. The household does not feel the need to have alternate system of power generation from solar, as the existing system is reliable enough. Many of the households without the rooftop solar system have stated that 'existing system is good enough' as one of the main reasons for not opting for solar power.

10.5 Conclusion and Recommendations

The central and state governments have devised many policies to promote rooftop solar power generation systems and further aid in the reduction of greenhouse gas emissions. These are now being integrated into building bye-laws and are a part of the environmental clearances required for the construction projects. The case study has shown that there are numerous opportunities and tremendous scope of solar energy in India. There is a need to further enhance these opportunities, in order to ensure stable and continual growth of solar power. The rooftop solar power generation in the residential sector is still at a nascent stage of development and has a few constraints, which need to be overcome, in order to maximize the true potential of this segment. Due to the lack of awareness related to economic and environmental potential of rooftops, residential consumers are still very reluctant to adopt this technology. There is a need to educate the consumers about the economic gains and the environmental benefits of adopting rooftop solar power generation systems.

The emphasis on economic gains and reduction in the electricity bill is going to be the main attraction for households to adopt rooftop solar systems. Even in a study conducted by Sommerfeld et al. (2017) in Queensland, Australia, economic benefits have been found to be the main motivation for installation of solar PV by residential consumers. Bank loans for this sector are currently available as part of 'home improvement loans' and not as soft loans. There is a need to include this sector under soft loans to push further growth. The households are aware that there is a possibility of adopting rooftop solar power generation, but they do not know the complete process, cost and the payback periods. The government needs to include these components in their awareness campaigns. The rooftop solar power generation is currently not attractive without subsidies, therefore the need to continue with these incentives. Another important point is that households rely more on the experience and recommendations of their friends and family to adopt these systems. Therefore, the government needs to involve those who have already adopted solar system as the ambassadors for promoting rooftop solar power generation.

References

Ajay L (2018) Amdavadis lead the pack in going solar. Ahmedabad Mirror, Ahmedabad
Dube AS, Bhirud NL, More KD (2017) Testing and performance assessment of 1 kWp solar rooftop system. Int J Emerg Trends Technol 4(3):9040–9043
Explorable Homepage (2018) Snowball sampling. https://explorable.com/snowball-sampling. Last Accessed 18 Mar 2018
Ezysolare (2016) Solar energy assessment report. Ahmedabad
Government of India (2011) Census of India 2011. Office of the Registrar General & Census Commissioner, New Delhi
Gujarat Electricity Regulatory Commission (GERC) (2017) Procurement of energy from renewable sources (Second Amendment) regulations 2017, Gandhinagar
Hagerman S, Jaramillo P, Morgan M (2015) Is rooftop solar PV at socket parity without subsidies? Energy Policy 89:84–94
Kappagantua R, Daniela A, Venkateshb M (2015) Analysis of rooftop solar PV system implementation barrier in Puducherry smart grid pilot project. SMART GRID Technol 21:490–497
Shivalkar R, Jadhav H, Deo P (2015) Feasibility study for the net metering implementation in rooftop solar PV installations across reliance energy consumers. In: International conference on circuit, power and computing technologies [ICCPCT], Nagercoil, India
Shukla AK, Sudhakar K, Baredara P, Mamat R (2017) Solar PV and BIPV system: barrier, challenges and policy recommendation in India. Renew Sustain Energy Rev 82(3):3314–3322
Sommerfeld J, Buys L, Vine D (2017) Residential consumers' experiences in the adoption and use of solar PV. Energy Policy 105:10–16
The Times of India (2018) Power transmission losses rise in Tamil Nadu. https://timesofindia. indiatimes.com/city/chennai/Power-transmission-losses-rise-in-Tamil-Nadu/articleshow/ 52392597.cms. Last Accessed 23 Apr 2018

Thomas R (2018) Constraints for rooftop solar power generation in urban areas (Unpublished Masters' Thesis). CEPT University, Ahmedabad

Tongsopit S, Moungchareon S, Aksornkij A, Potisat T (2016) Business models and financing options for a rapid scale-up of rooftop solar power systems in Thailand. Energy Policy 95:447–457

Torrent Power Limited (2017) Standard of performance compliance report 2010–2017. Gujarat Electricity Regulatory Commission, Gandhinagar

Town and Country Planning Organization (2016) Model building bye-laws. Ministry of Urban Development, New Delhi

Chapter 11
Renewable Energy in India: Policies to Reduce Greenhouse Gas Emissions

Neeru Bansal, V. K. Srivastava and Juzer Kheraluwala

Abstract The demand for energy is rising continuously, and the reasons are— economic development, urbanization, rising standard of living and increasing population. In India, 65% of the energy demand is currently met through power generated from fossil fuels, especially from coal-based power plants. These power plants are the highest contributor to the total greenhouse gas emissions in the country. The share of renewable energy is only 15%, excluding the hydropower. There are many countries in the developed world, which are aiming to become 100% reliant on renewable energy. Indian government is aiming to increase the contribution of renewable energy to honour the country's commitment to reducing intensity of greenhouse gas emissions. Solar energy is being considered as one of the main components of the renewable energy basket as the country receives 300 days of good solar radiation. This chapter focusses on the global discussion to reduce greenhouse gas emissions followed by the policies of the central government and the state governments in India to promote renewable energy, especially solar energy, to reduce its greenhouse gas emissions. There are some states, which have actively engaged in renewable energy initiatives, and an analysis of the policies of these states is highlighted. The country-specific policy initiatives for those having a major share of renewable energy have been included in the chapter.

Keywords Greenhouse gas emissions · Renewable energy · Solar energy
Policies

N. Bansal (✉)
CEPT University, Ahmedabad, Gujarat, India
e-mail: bansal.neeru@cept.ac.in

V. K. Srivastava
Sankalchand Patel University, Visnagar, Gujarat, India
e-mail: drvks9@gmail.com

J. Kheraluwala
Ernst & Young LLP, New Delhi, India
e-mail: juzer50@gmail.com

© Springer Nature Singapore Pte Ltd. 2019
N. Shurpali et al. (eds.), *Greenhouse Gas Emissions*, Energy, Environment, and Sustainability, https://doi.org/10.1007/978-981-13-3272-2_11

11.1 Introduction

The energy demand across globe has been rising, as it happens to be the prime driver for development. The reasons for escalating demand are economic development, population growth, urbanization and an increasing standard of living. The existing demand for energy is mainly met by fossil fuel-based power generation. Pollution and release of greenhouse gases (GHGs) and the resultant climate change are major environmental concerns associated with the conventional fossil fuel-based power generation. The issue of climate change is gaining global attention and is at the centre stage of global discussion. The effort is to produce low-carbon or carbon-free energy. This becomes especially important in the light of the Paris Agreement, 2015 where nations have committed to reducing their carbon footprint. This has made discussion on renewable energy, an important one, as it is the key to the sustainable energy future. The policies of the Indian government are aiming to increase the share of renewable energy to honour the country's commitment to reduce intensity of GHG emissions. Solar energy is being considered as a major alternative to generate renewable energy due to the country's geographic location as many parts of the country receive 300 days of solar radiation.

This chapter focusses on the renewable energy initiatives at global level to reduce GHG emissions, followed by the policies of the central government and the state governments in India to promote renewable energy, especially solar energy. The policy initiatives of the states in India which have actively engaged in renewable energy initiatives have been discussed. The policy initiatives by other countries having a major share of renewable energy have been included in the chapter.

11.2 Global Initiatives on Climate Change

11.2.1 The Paris Agreement, 2015

The Paris Agreement builds upon the United Nations Framework Convention on Climate Change (UNFCCC). On 12 December 2015, 195 countries assembled in Paris to chart out the world's first universal climate treaty, the Paris Agreement, aimed at mitigating the effects of rising global temperatures. It is the first comprehensive agreement on climate change, bringing all the nations towards a common cause. Under Kyoto, developing countries were not required to reduce their emissions, but in the Paris Agreement, even they have been made a party and are required to undertake and publish targets for a less carbon-intensive development. The Paris Agreement is officially ratified by 144 countries and entered into force in record time in November 2016. The treaty aims at cutting emissions and keeping global temperatures from rising more than 2 °C above pre-industrial levels, while striving to limit them to 1.5 °C. The treaty aims to strengthen the capacity of all parties to deal with the impact of global warming; to initiate a modern and superior technology and an enhanced capacity building framework, to provide transparency of support and action through a more vigorous transparency framework.

In order to help developing countries, developed nations are supposed to contribute $100 billion annually to developing countries by 2020. This would help the poorer countries combat climate change and foster greener economies. The individual countries are tasked with preparing, maintaining and publishing their own GHG reduction targets. The targets are to be revised every 5 years. The Paris Agreement has set a target of achieving a carbon-neutral world sometime after 2050 but not later than 2100. This entails a commitment to limiting the amount of GHGs emitted by human activity to the levels that trees, soil and oceans can absorb naturally.

Following the announcement by the President of the USA in 2017 that the USA will cease to participate in the Paris Agreement, the confidence of the parties to the Paris Agreement has suffered a setback. India has still taken a bold decision to spearhead the campaign against global warming and climate change. India being the third largest emitter of greenhouse gases has pledged to follow a low-carbon footprint model of development.

11.2.2 India's Commitment to the Paris Agreement

India has committed to reducing its carbon footprint by 33–35% from its 2005 levels by 2030. This means that the country will have to adopt multi-prolonged policies. Simultaneously, the country will have to transform the power sector, in order to shift the current energy sources more significantly towards renewable energy to reduce volumes of emissions per unit of gross domestic product (GDP), as 70% of GHG emissions in the country are from energy sector (Government of India 2015a).

11.3 Global Renewable Energy Scenario

The summits of the Parties to the Convention (COP)[1] over the past 25 years have resulted in strong commitments towards carbon-free or low-carbon energy generation. The boom in renewable energy began in Europe and Asia with Germany, France, Italy and Japan leading the way in the early 2000s. Their extensive commitment to developing renewable energy, especially solar power, has resulted in these countries having the highest installed solar capacity. The next stage of solar power development came after 2006, where Australia, China, the USA, the UK

---

[1]The 197 countries that have ratified the UNFCCC are called Parties to the Convention (COP).

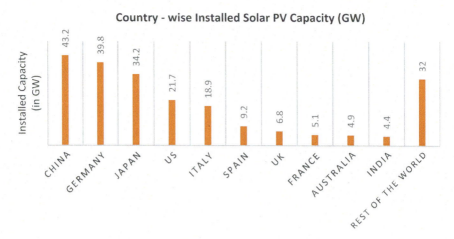

Fig. 11.1 Country-wise installed solar PV capacity (GW). *Source* IEA, 2017 based on 2015 data

began their advent in solar power generation. The top five countries in solar PV in 2016 are—China, Japan, Germany, the USA and Italy, respectively. China is the leader with a world share of 25.8% and has a total installed capacity of 78,100 MW. The details for the countries having a major share in the solar PV are shown in Fig. 11.1 (Hill 2017).

The country-specific policy initiatives by the major leaders in solar power generation are discussed in the following paragraphs.

11.4 Country-Specific Initiatives for Renewable Energy

11.4.1 China

China is the front runner in installed capacity of solar PV systems in the world. The Renewable Energy Law of China was formulated in 2006 and amended in December 2009. The country has been able to install more than the targets every year. The country used to have almost negligible solar PV systems till 2010, but since then, it has started setting massive installations every year.

The total installed capacity has increased from 2 GW in 2011 to around 130 GW in 2017 (Fig. 11.2). In the year 2017, it has added a massive capacity of 52.4 GW of solar PV systems. The policy follows a mechanism of feed-in tariff (FIT). This has resulted in tremendous annual growth every year in solar PV capacity addition. After witnessing the growth in 2017, the Chinese government has reduced the FIT by 12–15% since January 2018. The Renewable Energy Law has divided the country into Class-1, Class-2 and Class-3 regions based on solar resources. The

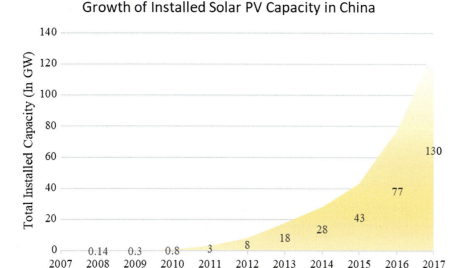

Fig. 11.2 Growth of installed solar PV capacity in China

revised FIT for 2018 is as follows: Class 1—CNY 0.55/kWh (Rs. 5.62); Class 2—CNY 0.65/kWh (Rs. 6.65) and Class 3—CNY 0.75/kWh (Rs. 7.67).

A direct visible impact of the push for renewable energy has resulted in reduced coal consumption in China since 2014. It is projected that renewable energy sources would contribute to 20% of China's total power consumption by the year 2030 (Greenpeace 2017).

11.4.2 Germany

The growth of the renewable energy sector in Germany started in the year 1991 under Electricity Feed Law (1991–2000). In 1991, the country initiated the "1000 solar rooftop" scheme. This scheme was followed by another rooftop solar PV programme, launched in 1999, known as the "100,000 solar rooftop programme". The programme provided financial assistance to systems greater than 1 kW at low interest rates with repayment options up to ten years. The programme targeted to install 300 MW by the year 2003 and was able to achieve 261 MW (Energiewende Team 2015).

During the "100,000 solar rooftop programme", in 2000, the Electricity Feed Law was superseded by the Renewable Energy Sources Act (EEG). The EEG has been quite effective since its inception. The period 2000 to 2004 saw an increase in the electricity generated from renewable sources from 1.48 to 3.98 GW. The share

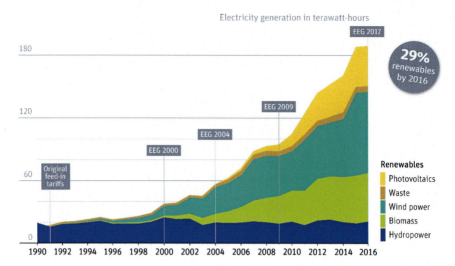

Fig. 11.3 Policy change and the growth of renewable energy in Germany

of wind and biomass increased two times and that of photovoltaic (PV) systems experienced a ninefold increase. In 2014, renewable energy contributed close to 30% of power generated in Germany with few peak days contributing close to 80% of peak power demand at specific times of the day (Morris and Pehnt 2016). As of 2017, the total power generation through solar PV systems in Germany is 43 GW, contributing to 7.2% of net energy consumption including grid losses.

Germany has set a yearly expansion plan of 2.4–2.6 GW up to 52 GW from solar PV systems. The energy generation from solar PV systems is largely driven by the residential sector. The major driver for increased installations has been gradual decline in the cost of solar PV systems. The latest amendment of the EEG Act in 2014 has shifted focus towards market-driven mechanism for fixing prices through bidding and auctions from the conventional system of FIT mechanism. This has resulted in reduction of feed-in tariff from 19 ct/kWh (Rs. 15) to a range of 9.2 to 13.1 ct/kWh (Rs. 7.37–Rs. 10.43), based on the size of the PV system. The timeline for the evolution of the Renewable Energy Law (EEG) in Germany from 2000 to 2017 along with the share of renewable energy from different sources is presented in Fig. 11.3 (https://www.iea.org/policiesandmeasures/pams/germany/name-21000-en.php).

11.4.3 Japan

Japan ranks at number three in terms of installed solar capacity. The country had invested heavily in research and development (R&D) for grid-connected solar rooftop programme under its Sunshine and New Sunshine Projects since the 1970s.

11 Renewable Energy in India: Policies to Reduce Greenhouse …

Table 11.1 Procurement prices for solar energy in 2012 (DLA Piper 2012)

Electricity generated	More than 10 kW	Under 10 kW	Under 10 kW (solar cogeneration)
Procurement price	JPY 42p/kWh	JPY 42p/kWh	JPY 34p/kWh
Procurement term	20 years	10 years	10 years

Japan had promoted solar energy by providing fiscal incentives to small generators through capital subsidy as well as by buying power through FIT mechanisms. The FIT mechanism being in force in Japan since 1 July 2012 had been the most generous one in the world. It mandated utility companies to purchase power from renewable energy generated at a price approved by Japan's Ministry of Economy, Trade and Industry (METI), at a rate shown in Table 11.1.

The country had set a modest target of 400 MW in 2000 and raised it to 4.6 GW by 2010. By 2016, the installed solar capacity in Japan was approximately 43 GW and accounted for 4.3% of the total electricity production. The Japanese government is currently projecting solar power generation to meet around 10% of total energy demand by 2050.

The Japanese FIT regime has been amended recently in April 2017 into the Renewable Energy Act. The main objectives of this Act are—achieve better growth of renewable projects, particularly solar; increase competition and improving the cost-effectiveness of renewable projects; promote gradual independence from the FIT regime; reduce renewable electricity costs for the public and introduce auction process for solar projects with a capacity greater than 2 MW. The intent of this auction system is to promote greater competition among solar developers which finally lead to a reduced electricity price for consumers. The shift from a FIT regime to a competitive auction process for large-scale solar projects is predictable, as these projects have attracted the bulk of investment in the renewable energy sector in Japan.

11.4.4 The USA

The USA has installed a total of 52 GW of solar energy as on 31 December 2017 as per the statistics released by the USA Energy Information Administration (EIA) in 2018. The solar energy accounts for almost 25% of all new power plants installed in the USA in 2017. Nearly 60% of installed solar capacity in the USA is in the form of large, utility-scale plants consisting of both solar PV and solar thermal. Similar to other countries, the national government has offered several incentives to promote and encourage the growth of solar power in the country. Coupled with this are the incentives offered by the respective state governments. Considering the massive size of the country and the federal nature of the government, each state has its own unique policy towards solar energy, although having certain common policy instruments. Some of the policy instruments are discussed in the following paragraph (http://theconversation.com/the-state-of-the-us-solar-industry-5-questions-answered-90578).

The Renewable Portfolio Standards (RPS) mandate utilities to purchase certain percentage of their power generation from renewable sources. Twenty-five out of the 50 states have mandatory RPS obligations to be met. However, there are 14 states in the USA which do not have RPS obligations; this includes Arizona which otherwise is in the list of top ten solar power states. Another important policy instrument in the USA is "net metering", and this has been adopted by nearly 37 states. A few states like Arkansas, California, Colorado, Louisiana do not have any restrictions on their net metering policies. The state of Arizona has recently restricted their net metering policy to only large-scale power projects and has recently excluded residential installations from the purview of this policy. Investment tax credit (ITC) also known as federal tax credits is another policy instrument to promote renewable energy. It is provided in the form of tax deductions equivalent to 30% of the total cost of the solar power system. This was introduced in 2005 and was initially promoted only for a period of two years. But after considering the positive response from consumers, it has been extended till 2021. Certain states have added additional tax credits to the ITC for installation of solar systems (https://www.solarpowerrocks.com/2017-state-solar-power-rankings/#propexemption).

11.4.5 New Entrant in Solar Energy—Saudi Arabia

The Kingdom of Saudi Arabia is the latest entrant in introducing renewable energy component in the conventional energy basket. The country which currently ships the largest quantity of crude oil in the world now plans to shift away from fossil fuel-based power. It has a yearly solar irradiance of more than 5.75 kWh/m^2 (Pazheri 2014). Saudi Arabia has launched the National Renewable Energy Programme (NREP) as part of Vision 2030 and National Transformation Programme (NTP). The NREP has two phases—0.45 GW of renewable energy generation by 2020 in phase 1 and a cumulative capacity of 9.5 GW by 2023 in phase 2. This will account for 10% of the country's total power requirement. Recently, in January 2018, Saudi Arabia has awarded 300 MW of utility-scale solar PV plant. There are plans to award an additional 620 MW by the end of 2018. However, rooftop solar PV projects in the country are not being given a major push; reason being the dusty climate conditions would demand higher maintenance costs for the small generators.

11.5 Power Sector and Emission of Greenhouse Gases in India

The energy demand in India is mainly met by fossil fuel-based power generation. In February 2018, India's total installed power capacity was 334 GW which includes 62.85 GW of renewable energy, and this makes the share of renewable energy to be approximately 15%, after excluding electricity generated from hydropower (Central Electricity Authority 2018). The installed capacity of thermal power plants, mostly coal-based plants, has continued to grow at a high growth rate of 16% during the period 2011 to 2017. However, in 2017, for the first time, the capacity addition of renewable energy was more than the conventional energy sources, wherein, 12.5 GW of renewable energy was added against 10.2 GW of conventional energy power.

The figures in Table 11.2 highlight that country is still majorly dependent on fossil fuels for meeting its energy requirements. Due to the nature of energy mix, almost 70% of the GHG (amounting to 1,510,120 Gg CO_2 equivalent) emitted into the atmosphere are from across the energy sector categories in 2010 (Government of India 2015a). India is the third largest GHG emitter in the world, though the per capita emissions are very small. The need is, therefore, to shift from conventional energy sources to new and renewable energy sources (Fig. 11.4).

In Paris Agreement, India has committed to reducing GHG emissions intensity by 33–35% of 2005 levels by 2030. Since energy sector is the major contributor to GHG emissions, many policy initiatives have been directed towards promoting renewable energy and specifically solar energy. These are discussed in the subsequent section.

Table 11.2 Total installed power capacity in India—2017 versus 2011

Source type	Type of generation	Total installed capacity (in GW)		% split	
		2017	2011	2017 (%)	2011 (%)
Conventional energy sources	Coal	192.97	93.84	58	54
	Nuclear	6.78	4.78	2	3
	Gas	25.15	17.46	8	10
	Diesel	0.84	0.00	–	–
Renewable energy sources	Hydro	49.35	40.32	15	23
	Wind	32.70	13.18	10	8
	Solar	14.77	0.03	4	0.02
	Biomass	8.30	2.67	3	2
Total installed capacity (in GW)		331	172	100	100

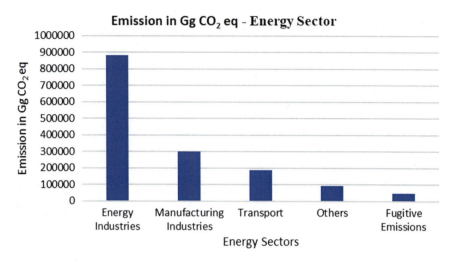

Fig. 11.4 Emission in Gg CO_2 eq from energy sector. *Source* Government of India (2015a)

11.6 National Action Plan for Climate Change (NAPCC) in India

The National Action Plan for Climate Change (NAPCC) was launched in the country in 2008 to tackle the issue of climate change. The objective of NAPCC is "to establish an effective, co-operative and equitable global approach based on the principles of common but differentiated responsibilities and respective capabilities". The NAPCC has formulated eight National Missions, which are fundamental to the National Action Plan for fulfilling the objectives set by the UNFCCC. These Missions are listed as follows.

1. National Solar Mission
2. National Mission on Enhanced Energy Efficiency
3. National Mission on Sustainable Habitat
4. National Water Mission
5. National Mission for Sustaining Himalayan Ecosystem
6. National Mission for Green India Mission
7. National Mission for Sustainable Agriculture and
8. National Mission on Strategic Knowledge for Climate Change.

11.6.1 Jawaharlal Nehru National Solar Mission (JNNSM)

Jawaharlal Nehru National Solar Mission (JNNSM) is one of the missions under NAPCC and was launched on 11 January 2011. The major objective of JNNSM is the development of solar energy; reason being that the country gets abundant sunlight for more than 300 days of the years. Solar is the most secure source of power as it is available in abundance throughout the country. The proximity to the equator also provides high solar radiation between 4 and 7 $kWh/m^2/day$. The country has a capacity to generate 5500 trillion kWh/year equivalent energy through solar radiation. This provides immense potential for development of solar power as an alternate source of renewable energy to wind, biomass and hydro-electricity. JNNSM has two segments for development of solar power, one through utility-scale projects and the other through rooftop solar PV projects.

Targets of the JNNSM. The mission is divided into three phases spanning from 2010 to 2022. The government will review and evaluate the capacity targets during the mid-term and end of each phase to prevent overexposure of government subsidies and vice versa. The targets for this mission were first set in 2010 and were revised by the cabinet in 2015. The targets under JNNSM are summarized in Table 11.3.

The installed solar capacity for power generation in 2009, prior to the National Solar Mission, stood at a meagre 0.02% (0.03 GW), which has now increased to 5%. The targets of the JNNSM have been revised from 20 to 100 GW of total solar installations by 2021–22. The development and commissioning of large-scale projects have been at the forefront of India's solar energy capacity addition. However, India lacks adequate transmission infrastructure to evacuate power from these large-scale power plants. Therefore, the generation of power at the point of consumption in urban areas has been given importance in the recent JNNSM policy amendment, through the implementation of rooftop solar PV systems. As per the revised government resolution of July 2015, the government aims to add 60 GW through medium and large-scale solar power projects like solar parks and another 40 GW of grid-connected solar rooftops on residential, commercial, institutional and industrial buildings (Ministry of New & Renewable Energy 2015; Goel 2016). In order to facilitate and promote the rooftop solar plants, Ministry of New and Renewable Energy (MNRE), India, has introduced another programme called Sustainable Rooftop Implementation for Solar Transfiguration of India (SRISTI), the details of which have been discussed in the next section.

Table 11.3 Targets of JNNSM

Phase	Year	Rooftop solar	Utility-scale solar plants	Total targets (in GW)		Target achieved
				2010 resolution	2015 revised resolution	Commissioned capacity (in GW)
I	2010–2013	No separate distinction		1.1	Not applicable	0.87 (PV, CSP, migration)
II	2013–2015	No separate distinction		3.0–10.0	Not applicable	3.74 up to 2015
	2015–16	0.20	1.8		2.0	3.0
	2016–17	4.80	7.2		12.0	2.89
III	2017–18	5.0	10.0	≥ 11.0	15.0	5.53 up to 30 Nov. 2017
	2018–19	6.0	10.0		16.0	
	2019–20	7.0	10.0		17.0	
	2020–21	8.0	9.5		17.5	
	2021–22	9.0	8.5		17.5	
Total revised target		40.0	57.0[a]	Not applicable	97.0[a]	
Total target of JNNSM (in GW)				20	100[b]	

[a]Around 3 MW of solar capacity was installed before the JNNSM was initiated
[b]After the Revision of JNNSM target to 100 GW in 2015, phase wise breakups of targeted capacity have not been released by the government

11.6.2 Sustainable Rooftop Implementation for Solar Transfiguration of India (SRISTI)

SRISTI programme has been proposed by the MNRE for implementation of rooftop solar PV plants. This programme focusses to ensure that the target of 40 GW from solar rooftops by 2022 is achieved smoothly (Government of India 2015b). The yearly targets for solar rooftops and the break-up based on the type of roof from various sectors like residential, commercial, government, industrial are given in Table 11.4.

The SRISTI programme has been formulated after considering the key learnings from the phase 1 of JNNSM. The prime focus is to reduce the complexities involved in the process of setting up rooftop solar systems. Accordingly, it has proposed to integrate the distribution companies (DISCOMs) as the primary implementation agency. The DISCOMs are likely to incur additional costs due to their responsibility towards capacity building, awareness programmes, etc. Therefore, performance-linked incentives will be provided to DISCOMs as a measure of compensation. It may be in the form of every MW capacity of solar energy added from rooftops (except for residential sector) into their distribution network. The residential sector is excluded as separate set of subsidies that have been announced for this sector. It is proposed that the funds linked to incentives will be released on a quarterly basis. This is an enabling policy instrument to ensure that

11 Renewable Energy in India: Policies to Reduce Greenhouse …

Table 11.4 Yearly targets for rooftop solar systems

Year	2015–16	2016–17	2017–18	2018–19	2019–20	2020–21	2021–22	Total
Target (in MW)	200	4,800	5000	6000	7,000	8,000	9,000	40,000
Sector-wise targets (SRISTI programme)								
Commercial and industrial sector	20,000 MW							
Government sector	5,000 MW							
Residential sector	5,000 MW		Central financial assistance available per kW					
Institutional sector	5,000 MW							
Social sector	5,000 MW							

Note The sector-wise targets maybe modified based on the demand in respective sectors

DISCOMs take an active part in promoting and implementing rooftop solar systems.

Central Financial Assistance (CFA) Scheme for residential sector: As per the provisions of the Central Financial Assistance (CFA) for the residential sector, the consumers in the residential sector shall be eligible for a 30% subsidy (CFA) per kW with a limit of 50 kW. The CFA will be released directly to the DISCOM either on a bimonthly or quarterly basis. An advance amount of up to 30% of the total CFA can be availed by government-owned DISCOM. In the case of private DISCOMs, the same will be provided on submission of a bank guarantee. This amount will then be passed on by the DISCOMs to the channel partners. The financial impact of the SRISTI programme is estimated at Rs. 23,450 crores to the exchequer.

11.7 State-Specific Solar Policies in India

In accordance with the JNNSM, certain states in India have drafted their own solar policies to promote the renewable energy. Based on the solar irradiance, the states of Gujarat, Tamil Nadu, Rajasthan have capitalized on this energy reserve with the support of the central government after the inception of JNNSM in January 2011. The solar power policy of these states is discussed in the following paragraphs. The states have incorporated subsidies, over and above the subsidy given by the MNRE. The states have also embedded other policy instruments like FIT and net metering to promote renewable energy. The following policies have been discussed:

- Tamil Nadu Solar Energy Policy, 2012
- Rajasthan Solar Energy Policy, 2014
- Gujarat Solar Power Policy, 2015.

11.7.1 Tamil Nadu Solar Energy Policy, 2012

Tamil Nadu has a high solar radiation of 5.6–6 kWh/m^2/day with around 300 sunny days a year. The aim of this policy is to install 3000 MW of solar power by 2015. The state has mandated renewable purchase obligation[2] (RPO) starting with 3% till December 2013 and 6% thereafter. The RPO will be administered by Tamil Nadu Generation and Distribution Corporation (TANGEDCO). A target of 3.5 GW by 2022 through rooftop solar PV systems for Tamil Nadu has been set by MNRE (Government of India 2015c). In order to promote rooftop solar PV systems, subsidies and/or incentives have been provided in the policy.

For the residential solar rooftops, a state subsidy of Rs. 20,000 per kW has been provided up to 1 kW plants. This is in addition to the subsidy by MNRE. The surplus power exported into the grid is adjusted over a period of twelve months. Settlement for surplus power is done only up to 90% of the total power consumed from the grid. Any power in excess of this value is considered void and not applicable for any payment. The subsidy provision is applicable only for residential consumers.

The rooftop solar systems installed on industrial and commercial buildings are not eligible for subsidy, but they are given generation-based incentives (GBI) at a rate of Rs. 2 per unit (first two years), Rs. 1 per unit (next two years) and Rs. 0.5 per unit (subsequent two years). For industrial consumers, demand cut exemption to the extent of 100% of the installed capacity assigned for captive use purpose is allowed.

The other enabling mechanisms in the policy are in terms of "single window clearance" guaranteed through TEDA[3] in 30 days, so as to enable commissioning of plants in less than a year. All the solar power developers/producers are also eligible to avail CDM benefits to enhance the viability of the projects. On account of the political change and uncertainty in Tamil Nadu since 2016, the state solar policy has not been revised after its expiry in 2015.

11.7.2 Rajasthan Solar Energy Policy, 2014

Rajasthan has the highest solar radiation in India, ranging between 6 and 7 kWh/m^2/day for more than 325 sunny days a year. The state has a potential to generate 142

---

[2]The Renewable Purchase Obligations (RPO) has been the major factor in India to promote the renewable energy sector. Under the RPO, states are supposed to achieve certain targets, specified by Central government, by ensuring that their power-share comes from green or renewable sources. In case the states are unable to produce enough renewable due to any-reasons, they buy Renewable Energy Certificates (REC) to compensate for the lag in the target. The State Electricity Regulatory Commissions (SERCs) define their respective RPO Regulations. The 'obligated entities' for RPO are mostly electricity distribution companies and large consumers of power.

[3]Tamil Nadu Electricity Development Authority.

GW as estimated by National Institute for Solar Energy (NISE). The state solar policy has set an ambitious target of 25,000 MW capacity installation by 2024. The total installed capacity of rooftop solar PV systems in the state crossed 129 MW in September 2017. A major factor which has affected the installation of rooftop systems in Rajasthan is the absence of any subsidy or incentive for the residential sector. Although Rajasthan provides the highest FIT rate (Rs. 3.93/unit) for export of surplus power to the grid, the cost of installation is a big deterrent for consumers. The MNRE has set a target of 2.3 GW by 2022 through rooftop solar PV systems for the state of Rajasthan.

11.7.3 Gujarat Solar Power Policy, 2015

Gujarat was the first state in India to develop a solar policy in 2009, two years before the commencement of JNNSM. This provided Gujarat a head start in generation of renewable energy. The learnings from the 2009 policy have helped to draft the revised policy in 2015. The policy has been formulated in accordance with the Electricity Act-2003 and shall be applicable till 31 March 2020 (Government of Gujarat 2015). Gujarat Energy Development Agency (GEDA) and Gujarat Power Corporation Limited (GPCL) have been the designated as the state nodal and facilitating agency, respectively.

Gujarat is rich in solar energy resources with a high solar radiation between 5.5 and 6 $kWh/m^2/day$. The state has a potential to generate around 10,000 MW of solar energy. The target set by MNRE for Gujarat for rooftop solar PV systems is 3.2 GW by 2022. The state solar policy has differentiated between projects for the purpose of subsidy and incentives. The different types of solar power projects as classified by the local DISCOMs are depicted in Fig. 11.5.

The policy has set eligibility criteria for availing subsidies and incentives for rooftop solar PV projects. The main highlights of the criteria are as follows:

- The project proponent shall own or be in legal possession of the premises including the rooftop or terrace.
- The project proponent shall be a consumer of the local DISCOM, and the premises shall be connected to the DISCOM's grid.
- The project proponent shall consume all the electricity generated from the solar PV system at the same premises. If unable to consume all the generated electricity, relevant provisions as defined below shall be applicable to the generated surplus electricity.

Subsidy, Incentives and Feed-in Tariffs. The government of Gujarat and the MNRE provide subsidy towards installation of grid-connected rooftop solar PV systems. The benchmark cost set by Gujarat Energy Development Agency (GEDA) for the installation of 1 kW solar PV system is Rs. 69,000 as per the mandated specifications. The specification includes all equipment costs, installation costs and

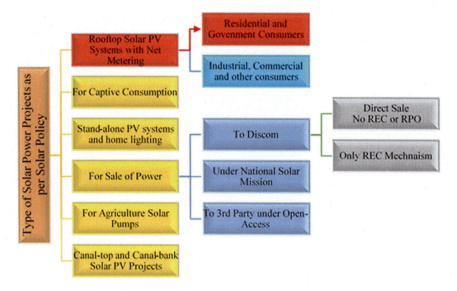

Fig. 11.5 Types of solar power projects in Gujarat. *Source* Adapted from Gujarat Solar Power Policy, 2015

the operation and maintenance (O&M) costs for 5 years. To the residential rooftop solar PV systems, the MNRE provides a subsidy at a rate of 30% of the total cost, amounting to Rs. 20,700 per kW, with a maximum limit of 50 kW. The state also provides an additional subsidy at a rate of Rs. 10,000 per kW, with a limit of 2 kW. The subsidy is directly provided to channel partners. This means that the residential consumer/beneficiary is required to pay only Rs. 38,300 per kW. The subsidies are not provided to commercial, industrial and institutional consumers by the state government and the central government. However, they are provided with the benefits of "accelerated depreciation".

In the case of net export of surplus power into the grid, from a residential rooftop solar PV systems, the DISCOM pays FIT to the household at a rate of Rs. 3.26 per kWh. For other rooftop solar PV systems, the FIT is higher at Rs. 6.61 per unit (in case of appreciated depreciation benefits) and Rs. 7.28 per unit (without appreciated benefits).

11.8 Conclusion

There is a global consensus on reducing the emission of GHGs and keeping the rise in temperature under check. The formation of policy and enabling instruments to promote clean energy will have a major impact in reducing pollution levels and decline in GHG emissions. Initiatives have been taken by many countries to promote renewable energy and reduce their carbon footprint. Worldwide experience indicates that rooftop solar PV installations have played a significant role in increasing the share of solar power installations in Europe and Japan. In 2011, Germany added 60% of all solar generation capacity through rooftop solar PV projects. Today, Germany has more than one million rooftop solar PV installations. Likewise, Japan has seen growth primarily through rooftop solar PV installations and more recently by integrating solar cells into the building façades, sometimes complementing and in many cases completely replacing traditional view or spandrel glass. Some of the key factors supporting the expansion of rooftop solar PV installations are—the sector's ability to thrive under conducive policy and enabling regulatory environments; the non-existent threat of supply of encumbrance free land and the ability of rooftop solar PV projects to meet directly the power requirement, thus reducing transmission and distribution losses and thereby avoiding the expensive infrastructure required for power evacuation.

The global solar PV market grew significantly, by adding around 74.4 GW in 2016 alone. The growth was in spite of the fact that the two major contributors, Japan and Europe, did not add any significant capacity. Asia is ranked first for the fourth consecutive year with around 67% of the global PV market in 2016, up from 60% in 2015. China alone added 52 GW of installed capacity in 2017, and outside of China, the global PV market grew between 35 and 40 GW (IEA 2018).

India has committed to reducing its carbon footprint and has taken initiatives at all levels to promote renewable energy. The renewable energy includes solar energy, wind energy, bioenergy, hydropower. The solar energy has been focused in the chapter; reason being that this is the main focus sector in the National Action Plan for Climate Change. The national policy on renewable energy has set a target of 175 GW of renewable energy generation by 2022, out of which 100 GW is to be generated as solar energy. The generation of hydropower and wind energy is possible only at specific locations, whereas solar energy is available in most parts of the country, with a good radiation intensity, for the most part of the year, due to the country's geographic location. The initiatives to promote solar energy have been taken at all levels—from the central government to the state governments, local DISCOMs and the individual household level. The end users across industrial, commercial and residential sectors have been involved in the process of generating solar energy by offering certain incentives. The incentives have been built as the markets alone cannot solve the problem of GHG emissions related to energy production. The government has supported the shift to renewable energy by providing incentives in the form of subsidies, tax rebates, etc. The proposals like the SRISTI programme are a step in the right direction towards the promotion of grid-connected rooftop solar systems. This new scheme is a step forward in involving the DISCOMs as active participants in the implementation of rooftop solar systems by providing incentives towards the achievement of targets fixed by the MNRE. The continued increase in capacity addition of renewable energy will help the country to meet its international obligation of reducing the emission of GHGs.

References

Central Electricity Authority (2017) All India installed capacity—monthly report: December 2017, p 3

Central Electricity Authority (2018) All India installed capacity—monthly report: February 2018, p 3

DLA Piper (2012) Japan's renewable energy feed-in-tariff regime, Brisbane

Energiewende Team (2015) How is Germany integrating and balancing renewable energy today? https://energytransition.org/2015/02/how-germany-integrates-renewable-energy/. Accessed 02 May 2018

Goel M (2016) Solar rooftop in India: policies, challenges and outlook. Green Energy Environ 1:129–137

Government of Gujarat (2015) Gujarat Solar Power Policy—2015. Gandhinagar

Government of India (2015a) First Biennial Update Report to the United Nations Framework Convention on Climate Change. Ministry of Environment, Forest and Climate Change, New Delhi, p 57

Government of India (2015b) SRISTI (Sustainable Rooftop Implementation for Solar Transfiguration of India)—318/331/2017-GCRT. Ministry of New & Renewable Energy, New Delhi

Government of India (2015c) Statewise and yearwise targets for installation of 40,000 MWp grid connected solar rooftop systems. Ministry of New & Renewable Energy, New Delhi

Greenpeace (2017) By 2030 China's wind and solar industry could replace fossil energy sources to the tune of 300 million tonnes of standard coal per year. http://www.greenpeace.org/eastasia/press/releases/climate-energy/2017/By-2030-Chinas-wind-and-solar-industry-could-replace-fossil-energy-sources-to-the-tune-of-300-million-tonnes-of-standard-coal-per-year/. Accessed 25 April 2018

Hill JS (2017) China officially installed 52.83 gigawatts worth of solar in 2017. https://cleantechnica.com/2018/01/22/china-officially-installed-52-83-gw-worth-solar-2017-nea/. Accessed 02 May 2018

International Energy Agency (IEA) (2018) Snapshot of global photovoltaic markets—photovoltaic power systems programme. https://www.researchgate.net/publication/324703156_2018_SNAPSHOT_OF_GLOBAL_PHOTOVOLTAIC_MARKETS. Accessed 02 May 2018

International Energy Agency (IEA) https://www.iea.org/policiesandmeasures/pams/germany/name-21000-en.php. Accessed 12 April 2018

International Energy Agency (IEA) (2017) Key world statistics

Ministry of New & Renewable Energy (2015) Resolution No.30/80/2014-15/NSM

Morris C, Pehnt M (2016) The German Energiewende book. Heinrich Böll Foundation, Berlin

Pazheri FR (2014) Solar power potential in Saudi Arabia. Int J Eng Res Appl 171–174 (2014)

Renewables surpass other energy sources in capacity addition in FY17. https://www.livemint.com/Industry/yWPo2ZUyvn4hZPmQhtORoM/Renewables-surpass-other-energy-sources-in-capacity-addition.html. Accessed 23 April 2018

The state of the US solar industry: 5 questions answered. http://theconversation.com/the-state-of-the-us-solar-industry-5-questions-answered-90578. Accessed 30 April 2018

United States solar power rankings (2017). https://www.solarpowerrocks.com/2017-state-solar-power-rankings/#propexemption. Accessed 30 April 2018